Care and Maintenance of Textile Products Including Apparel and Protective Clothing

TEXTILE INSTITUTE PROFESSIONAL PUBLICATIONS

PUBLISHED TITLES

Care and Maintenance of Textile Products Including Apparel and Protective Clothing
Rajkishore Nayak and Saminathan Ratnapandian

Care and Maintenance of Textile Products Including Apparel and Protective Clothing

Rajkishore Nayak and
Saminathan Ratnapandian

CRC Press
Taylor & Francis Group
Boca Raton London New York

CRC Press is an imprint of the
Taylor & Francis Group, an **informa** business

The Textile Institute

Contents

Series preface

Textile Institute Professional Publications

The aim of the **Textile Institute Professional Publications** is to provide support to textile professionals in their work and to help emerging professionals, such as final year or masters students, by providing the information needed to gain a sound understanding of key and emerging topics relating to textile, clothing and footwear technology, textile chemistry, materials science and engineering. The books are written by experienced authors with expertise in the topic and all texts are independently reviewed by textile professionals or textile academics.

The textile industry has a history of being both an innovator and an early adopter of a wide variety of technologies. There are textile businesses of some kind operating across the world. At any one time, there is an enormous breadth of sophistication in how such companies might function. In some places where the industry serves only its own local market, design, development and production may continue to be based on traditional techniques; but companies that aspire to operate globally find themselves in an intensely competitive environment, some driven by the need to appeal to followers of fast-moving fashion, others by demands for high performance and unprecedented levels of reliability. Textile professionals working within such organisations are subjected to a continued pressing need to introduce new materials and technologies, not only to improve production efficiency and reduce costs, but also to enhance the attractiveness and performance of their existing products and to bring new products into being. As a consequence, textile academics and professionals find themselves having to continuously improve their understanding of a wide range of new materials and emerging technologies to keep pace with their competitors.

The Textile Institute was formed in 1910 to provide professional support to textile practitioners and academics undertaking research and teaching in the field of textiles. The Institute quickly established itself as the professional body for textiles worldwide and now has individual and corporate members in over 80 countries. The Institute works to provide

sources of reliable and up-to-date information to support textile professionals through its research journals, the *Journal of the Textile Institute* [1]
and *Textile Progress* [2], definitive descriptions of textiles and their components through its online publication *Textile Terms and Definitions* [3] and
contextual treatments of important topics within the field of textiles in
the form of self-contained books such as the *Textile Institute Professional
Publications.*

References

1. http://www.tandfonline.com/action/journalInformation?show=aimsScope
   &journalCode=tjti20
2. http://www.tandfonline.com/action/journalInformation?show=aimsScope
   &journalCode=ttpr20
3. http://www.ttandd.org

Authors

Dr. Rajkishore Nayak is currently working as a senior lecturer at the School of Communication and Design, RMIT University, Vietnam. He completed his PhD from the School of Fashion and Textiles, RMIT University, Australia. He has 15 years of experience in teaching and research related to fashion and textiles. He has published about 100 peer-reviewed papers in national and international journals. Rajkishore was awarded with the 2015 RMIT University Research Excellence Award. He also received the 2012 RMIT University Teaching and Research Excellence Award and 2008 RMIT University International Scholarship. He worked with the School of Fashion and Textiles, RMIT University, Australia from 2012–2016 in teaching and research.

Dr. Saminathan Ratnapandian is a professor at the Ethiopian Institute of Textile and Fashion Technology, Bahir Dar University, Ethiopia. He earned his PhD from the School of Fashion and Textiles, RMIT University, Australia in 2013. He was a research fellow at TRI/Princeton (masters degree) and RMIT University (doctoral degree). His publications are available in a broad spectrum of reputed journals related to fashion and textiles. He has served the sector for nearly 25 years.

List of abbreviations

AATCC	American Association of Textile Chemists and Colorists
AOX	adsorbable organo-halogen
AS/NZS	Australia and New Zealand standard
ASTM	American Society for Testing and Materials
BOD	biochemical-oxygen demand
CARB	California Air Resources Board
CEN	Comite Europeen de normalization
COD	chemical-oxygen demand
DfE	Design for the Environment
DFE	directional frictional effect
DP	durable press
DPTB	dipropylene glycol tertiary butyl ether
DWR	durable water repellent
EPA	Environmental Protection Agency
FFPPC	firefighter's personal protective clothing
FR	flame retardant
FRPPC	flame retardant personal protective clothing
FSP	fragment simulating projectile
FTC	Federal Trade Commission
FTIR	Fourier-transform infrared
GEC	GreenEarth Cleaning
ISO	International Organization for Standardization
JIS	Japan Industrial Standard
LAS	linear alkyl benzene sulfonates
NAFTA	North American Free Trade Agreement
NFPA	National Fire Protection Association
NIJ	National Institute of Justice
NPE	nonylphenol ethoxylates
OBA	optical brightening agents
P/C	polyester/cotton
PAC	polyacrylic
PBO	piperonyl butoxide
PCE (perc)	perchloroethylene

PET	polyethylene terephthalate
PLA	polylactic acid
PLC	programmed logic circuit
PPC	personal protective clothing
PU	polyurethane
PVC	polyvinyl chloride
REACH	Registration, Evaluation, Authorisation and Restriction of Chemicals
RFID	radio frequency identification
RH	relative humidity
SPF	sun protection factor
UBACS	Under Body Armor Combat Shirt
UHMWPE	ultra-high molecular weight poly ethylene
UV	ultraviolet
VOC	volatile organic compound

chapter one

Introduction

Textile products get soiled, stained, dirty and even worn during their use, and may not be usable after a certain period [1,2]. Hence, they need regular care and maintenance, which helps to extend the durability of the clothing and ensures that fresh clothing is ready to wear when needed [3]. However, this is one of the most neglected aspects by many consumers. Most garments cannot be put into the washing machine straight and come out perfect after washing. Hence, proper care is necessary to retain the original properties. Proper care and maintenance helps to reduce the budget allocated to purchasing clothes and improve the wearability. Furthermore, this can reduce the environmental impact by the reduction in the usage of raw materials, processing chemicals and power consumption [4].

On one hand, the technological developments in the washing machine and detergency have reduced the total environmental impact per wash. On the other hand, the amount of clothing owned by individuals has gone up, which has also lead to the increased frequency of washing. It is worth mentioning that during the use of a specific garment, the most energy is consumed in its cleaning and maintenance [5]. Hence, the cleaning and maintenance of the textiles in the right time with appropriate chemicals and protocols cannot only reduce their environmental impact but also improve their durability [6].

A care label carries instructions for the cleaning of a textile product [2,4,7–9]. Care labels contain a series of directions describing procedures for refurbishing a product without adverse effects. Care labelling for garments is essential to identify the product, to assist the consumer in product selection and the retailer in selling the product, and to help the consumer in effective care of the garment [10]. The information on care labels is strongly emphasised as most consumer complaints and claims against apparel products concern colour change, deformation and damage during laundering.

Manufacturers of textile items provide proper care instructions in the clothing. However, the inability to follow the instructions, to select appropriate chemicals or washing cycles lead to permanent damage to the clothing [11,12]. It is the consumer's responsibility to take proper care of the textiles [13]. Most consumers who take care of the textiles might have experienced one or more problems such as colour fading, shrinkage, wrinkling and damage to the buttons, bids and sequins or other mechanical damage [14–20].

Several factors such as the type of fibre, type of detergent or chemicals used, temperature, agitation and duration affect the clothing properties after washing or dry cleaning [21–23]. The wrong selection of any of these parameters or in combination can damage the whole garment. Hence, the care labels need to always be followed before selecting any washing cycle or chemicals for the cleaning of the clothes.

This book covers the methods of cleaning clothes, namely wet cleaning or laundering and dry cleaning, which are most commonly used for the care of clothing items. The chemicals and types of machines used for these processes will also be discussed along with their ecological concerns. In addition, the potential new eco-friendly chemicals currently being used in these processes will be highlighted. Furthermore, the types of stains, their removal and impact of stain removal on clothing properties will be covered. The care and maintenance of woollen clothing items will also be discussed.

chapter two

Cleaning of textile materials

The history of cleaning of clothes or laundering dates back to 2500 BC, when soap was used in washing around the Tigris and Euphrates Rivers (now southern Iraq) [24]. Laundering is the process of removing stains, dirt, bad smells and microbes from clothes usually using water as a solvent, containing detergents or other chemicals accompanied by agitation, heat and rinsing. In some instances, cleaning uses solvents other than water and other specialty chemicals and equipment. The former process using water is known as washing and the latter process using chemical solvents is called dry cleaning.

The effectiveness of laundering clothes depends on the kind, amount and temperature of water; soaps, laundry aids and detergents [25–29]. The hardness of water, turbidity, colour, dissolved salts and metals may also affect the laundering [30–32]. The degree of soil removal in the cleaning process depends on the fabric substrate, fibre geometry, yarn and fabric structure, type of soil, chemical finish, cleaning method and interaction among all these factors [33–36]. Fibres with polar surfaces such as cotton and rayon can interact very strongly with water, whereas hydrophobic fibres such as polyester have been shown to interact with water slowly by dispersion forces, which may affect the cleaning efficiency.

Fibre geometry such as fibre diameter, cross-section, surface contour and crimp can affect the soil-retention property [37]. Fibres with a larger diameter, circular cross-section and smooth surface are not easily soiled. The presence of surface irregularities in the fibre acts as a sink for deposition of soil. During laundering, soil release from these sinks is a much slower process compared to a smooth surface. Increased mechanical action and/or increased detergent concentration can improve the soil removal from these areas.

The yarn structure involves staple or filament yarn, yarn fineness (coarse and fine) and amount of twist in the yarn [38,39]. High twist and staple yarns may prevent soil removal during laundering. Fabric structure involves woven, knits, nonwovens and composites, which may also vary in tightness, weight and thickness [40]. Open structures can offer less resistance to soil removal compared to tight structures. Similarly, lightweight fabrics can be cleaned more easily than their heavier counterparts.

The presence of various functional finishes in the fabric can alter the soil-removal efficiency from clothing. For example, the durable press (DP) finish can reduce the soil removal from cotton fabric. Similarly, the application of soil-release or soil-repellent finishes to fabric can improve the efficiency of cleaning. Other finishes such as flame retardant, hydrophilic/hydrophobic, antimicrobial [41,42], finishes to improve the handle and comfort can alter the cleaning efficiency depending on the nature of both the finish and the substrate.

Soil release from a textile material can involve three consecutive steps such as: (1) the induction phase, when the water and surfactant get diffused into the soil-fibre interface and into the soil, (2) separation of soil from the fabric and (3) final phase (leveling), when the soil removal is very slow [43–45]. The constituent of soils may involve solid particles; liquids such as oils; and mixtures of both solids and liquids. The surface tension of oily soils is significantly lower, which can penetrate fibres more readily than water-based soils. The viscosity of the oil is a deciding factor in soil release. The higher the viscosity, the harder it is to remove from the substrate. Oily soils can be more easily removed from synthetic fibres such as polyester than from cotton.

Clay soils adhere firmly to textile fibres as they are small, have an active surface, behave as colloids and have a large surface area in proportion to mass. Soil particles are more deposited at sites where a geometric bond is formed either in a fibre crevice or an interstitial void between fibres and yarns. The ease of their removal depends on the nature of the washing solution and the mechanical energy during the laundering process.

2.1   Wet cleaning (using water)

Wet cleaning or washing is the method of cleaning clothes that is usually done with water, often in the presence of a soap or detergent. Soaps and detergents are used for the emulsification of oils and dirt particles so that they can be easily washed away [46,47]. The washing will often be done at a temperature above room temperature to increase the activities of any chemicals used and the solubility of stains. In addition, high temperature kills microbes that may be present on the fabric.

Laundering is a complex process as an improper selection of parameters such as temperature, agitation, time and chemicals can permanently damage the fabric [48]. Furthermore, if the fabric is being treated with any of the functional finishes such as waterproofing, flame retardant, permanent press, deodorizing, antibacterial, soil release and pest control, the washing behaviour will be different. Special techniques should be adopted to improve the durability of the finishes.

Laundering is an exercise to decontaminate clothes that were in contact with the body, as the body is a source of contaminating the clothes [49]. This helps in clothing care and restores the clothing attributes such as style, feel and appearance. The laundering process provides freshness to the clothing items that are soiled, stained and musty, which make them ready to wear [50]. Laundering is always considered as an assembly of mechanisms consisting of textiles, detergents, washing machines and skills. The recent trend in laundering is more frequent washing at lower temperatures as compared to the less-frequent washing discussed earlier.

In addition to the factors related to the machine and chemicals, the hardness of the water can also affect the quality of washing [30,31,51,52]. The soaps and detergents are less effective in hard water. To avoid this problem, approaches such as the use of more soap or detergent, a longer washing cycle or higher temperature is necessary. Washing parameters such as frequency, washing temperature, type of detergent, use of a tumble dryer and ironing conditions are often related to the culture [53,54]. For example, cotton T-shirts are washed in cold water by Spanish consumers (48%), whereas Norwegians (48%) prefer to wash the same products at 60°C [53]. Similarly, the average washing temperature in Europe is about 45.8°C [55].

A comparative study of energy and water consumption of automated laundering around the globe showed that the energy use per wash cycle mainly depends on the average washing temperature [54]. An estimated value for lowering the washing temperature and eliminating both tumble-drying and ironing of a cotton T-shirt can lead to around a 50% reduction in global climate impact [56].

2.1.1 Washing with a machine

While laundering in a machine, it is essential to understand the nature of clothing, equipment to be used, chemicals used and when to launder [57]. Two important things, namely the colour of the clothes and the material they are made of, should always be considered while washing. Generally, for washing purposes, the colour of the clothes can be considered as 'light' or 'white' and 'dark', which should not be washed together. The material specification of textile items is mostly indicated on the care label. It is essential to consider the material type before washing as the washing protocols and laundering chemicals can permanently damage the cloth.

Light- and dark-coloured items should always be separated during washing. Light-coloured items should not be washed with dark-coloured items to avoid the risk of cross staining [58,59]. While washing new clothes, the dyes can leach from the fabric and stain other clothes in the load. Hence, the new clothes should always be washed alone if a machine is used or washed separately by hand. Colours such as white, cream or a pale pastel shade should be considered as 'whites' and should be washed separately from the other dark colours.

The care instructions should be always followed while washing, drying and pressing the clothes [60]. Some clothes need to be only washed by hand or dry-cleaned and dried flat or dried under shade. Deviation from these specifications may cause colour fading or the change of the size and shape. Hence, following the care instructions can help to clean the clothes appropriately, to retain the aesthetics, dimensions and to increase the wearability.

All fabrics cannot be washed with the same washing and tumbling conditions. For example, denim fabrics or terry towels need to be washed in a heavier cycle than the inner clothing or delicate clothing items. Hence, these items should be separated and washed in different loads to avoid any potential damage to the cloth. This will also ensure that all the clothes in a particular laundry load are properly cleaned.

The types of washing machines (i.e., top-and front-loading) are described in Section 3.1. The top-loading machine is better for the thicker fabrics, whereas the front-loading machine is preferred for thinner and delicate fabrics as it is less harsh [54]. Hot water should be used for light colours of cellulosic fabrics and their blends that are comparatively dirtier or stained, and can help in the easy removal of dirt and stains. However, if the cloth looks cleaner, cold water can be used. Similarly, dark-coloured items should be washed with cold to lukewarm water, which helps to retain the original shade. Hot water can remove the colour from these items.

While selecting the washing cycle in a washing machine, it is essential to understand the parameters such as time, temperature and agitation for each cycle. Different kinds of clothing need different washing cycles. The regular or normal cycle is selected for light colours, which will help the whites to retain their original colour and crispness. The delicate cycle can be used for washing relatively delicate clothing such as woollens, knits, silk items, bras, cotton sweaters and shirts. Delicate items should be always checked and confirmed that they are not meant for hand cleaning or dry cleaning. Permanent press should be selected for dark colours as it uses warmer water initially and cooler water at the end, which helps to retain the brightness of colours.

The selected washing load should be small, medium or large depending on the amount of clothes and the machine capacity. For example, a load that occupies about one-third of the space is considered as small; a two-thirds load is considered as medium; and a full load is considered as large. It is imperative to go for another washing cycle if the load exceeds the maximum capacity of the washing machine. Otherwise, it can lead to the risk of damaging or jamming the washing machine and/or improper cleaning with residual chemicals.

Washing chemicals and other washing aids, e.g., bleaches and softeners (see Sections 2.1.3 and 2.1.4), should also be selected appropriately. The amount of detergent should be selected on the basis of the load and degree of the soiling of clothes. The use of excessive chemicals can leave traces even after a complete washing cycle. This may be proven to be detrimental to the fabric or can cause irritations to the skin of some wearers [61–64]. The strength or concentration and nature of detergents commercially available vary a lot. Hence, the instructions on the detergent package and the nature of the cloth should always be considered before washing. Generally, softeners are added during the rinsing cycle. Washing machines have dispensers to add the softener at the start of the cycle, which is automatically added to the clothes during the appropriate rinse cycle.

While using a dryer for drying the clothes, the items specifying 'do not tumble dry' should be dried in shade or sunlight. The drying temperature and time are the parameters that need to be selected carefully. The drying cycle can be classified as regular, permanent press and delicate. The regular cycle should be selected for whites as they can handle heat better than the coloured items. The permanent press is less severe than the regular and should be used for coloured clothes. The clothes washed in the delicate cycle of the washing machine should be dried in the delicate cycle of the drier as well. The delicate cycle uses air at near room temperature and a slow cycle to prevent damage to the clothes.

While drying the clothes in sunlight or shade, good and strong hangers are essential to support the garments during drying. As the weight of a garment increases after washing, the hangers may break or bend. If unnoticed, the garment can lose its shape during drying. Some clothing, especially knits, can change dimension if hang dried. These items should be dried flat as described in the care label.

2.1.2   Washing by hand

The clothes labelled as 'Hand wash only' should be washed by hand. A bucket or a plugged sink can be used for hand washing. Cold or lukewarm water should be added first to the bucket followed by adding the suitable detergent [65]. The detergent should be mixed thoroughly by stirring by hand. The detergents used for hand washing are generally different from the detergents used in machine washings [66]. These detergents also vary in the concentration and nature, hence, should be carefully selected. The clothes should be dipped into the bucket and swished so that they are completely soaked with water. The bucket should be left for about 30 minutes so that the clothes are saturated with the detergent. Then the clothes should be rinsed with lukewarm and clean water at least twice or more so that the detergent is completely removed from the clothes. The items hand washed should not be hang dried as it can cause stretching of the fabrics. They should be dried flat to retain the shape and minimise the amount of wrinkles formed during the drying process.

The care labels work as a major source of information on the clothing type, type of washing and drying and nature of chemicals suitable for the same [67–69]. Hence, they should be always referred to before the clothes are washed. The care instructions are generally provided by symbols or words or by the combination of both. If it is hard to understand them, the commercial organisation or the manufacturer or seller can be contacted for getting the appropriate information. The following section (Figure 2.1 and Table 2.1) briefly describes the information on the care instructions

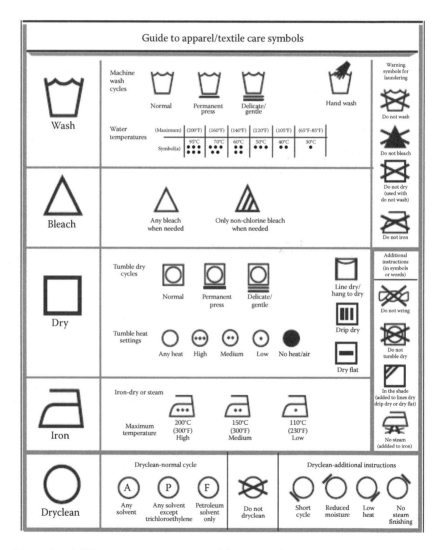

Figure 2.1 ASTM care instructions used for apparel (symbols).

Table 2.1 Care instructions used in apparel (words)

Specification on the care label	Instruction for proper care
Machine washing	
Machine wash	Washing by machine in lukewarm water. May be bleached but no dry cleaning.
Home wash only	Same as above, but no commercial laundering.
Cold wash/rinse	Wash and rinse with cold water from the tap.
Warm wash/rinse	Wash and rinse with warm water from the tap.
Hot wash	Use hot water machine setting.
No bleach	Avoid the use of bleach.
No starch	Avoid the use of starch.
Delicate/gentle cycle	Wash with appropriate conditions or wash by hand.
Durable/permanent press cycle	Wash with appropriate conditions for washing; otherwise use medium wash, cold rinse and short spin cycle.
Wash separately	Wash alone or with similar colours.
No spin	Do not apply the spin cycle after washing.
Hand washing	
Hand wash	Wash by hand using cold or lukewarm water. May be dry cleaned or bleached.
Hand wash only	Wash by hand in lukewarm water. May be bleached but no dry cleaning.
Hand wash separately	Wash by hand alone or with similar colours.
Damp wipe	Surface clean only with a damp cloth or sponge.
No bleach	Avoid the use of bleach.
Home drying	
Tumble dry	Tumble dry with no, low, medium or high heat.
Tumble dry/remove promptly	Same as above. Remove the clothes promptly when the tumbling stops.
Dry flat	Dry on a flat surface.
Line dry	Hang damp for drying.
Drip dry	Hang wet for drying with hand-shaping only.
No wring/no twist	Avoid wringing. Hang dry, drip dry or dry flat only. Handle carefully to avoid wrinkles or distortion.
Block to dry	Maintain original shape and size during drying.

(Continued)

Table 2.1 (Continued) Care instructions used in apparel (words)

Specification on the care label	Instruction for proper care
Ironing/pressing	
Do not iron	Avoid ironing.
Cool iron	Set iron at the lowest temperature.
Warm iron	Set iron at the medium temperature.
Hot iron	Set iron at the hot temperature.
Iron damp	Dampen clothes before ironing.
Steam iron	Iron or press with steam.
Miscellaneous	
Dry clean only	The garment should be dry cleaned only, including the self service.
Professional dry clean only	Avoid dry cleaning by self service.
No dry clean	Do not dry clean or use dry cleaning chemicals. Follow the care instructions.

Source: Kefgen, M. and Touchie-Specht, P., *Individuality in Clothing Selection and Personal Appearance, A Guide for the Consumer*, 3rd edition, Macmillan Publishing Co., Inc., New York, 1981.

mainly used in apparel clothing both for machine and hand washing. The detailed care labelling instructions are given in Chapter 4.

2.1.3 Washing chemicals

Laundering chemicals (soaps and detergents) are added to water to lower the surface tension for the ease of cleaning [46,70–72]. The laundering chemicals are available as powder, liquid, spray or granules. Soaps are metallic salts (aluminium, sodium, potassium) of fatty acids and are soluble in water. The soap molecule has two distinct parts: a carboxylate group (attracted to water) and a hydrocarbon chain (repelled by water). On the other hand, a detergent is a chemical composition that removes soiling and is produced by chemical synthesis.

Although both soaps and detergents are surfactants (or surface-active agents), they are not the same [73–75]. Soaps are usually manufactured from natural materials while detergents are made from synthetic materials [76]. Although soaps were the first detergents, they are now being replaced by synthetic detergents. Soap is highly deactivated by hard water. At the early stage of development of non-soap surfactants, the term *syndet* (short for synthetic detergent) was used to indicate the distinction from natural soap [77].

Synthetic detergents may be classified as anionic, cationic and non-ionic [78]. Anionic detergents are so-called because the detergent portion of the molecule is an anion (negative ion) and the water-soluble portion is a cation (positive ion). Most of the synthetic detergents commonly used in laundering are of the anionic type in which linear alkyl benzene sulfonates (LAS) are the main anionic compounds [79,80]. In cationic detergents, the detergent portion is cationic and the water-soluble portion is anionic. The cleansing action of cationic detergents is weaker than that of most anionic detergents. They are used as domestic germicides and fabric softeners. Non-ionic detergents are electrically neutral, having a neutral pH and are not affected by acids, alkalis or hard water. These detergents are very similar to other detergents in having one part of their molecule water-soluble and another part solvent-soluble. Some non-ionic detergents clean well and have very little lathering action in water. All three types of detergent molecules are explained in Figure 2.2.

Synthetic detergents consist of different components perform different functions. The two major components of detergents are surfactant and builder [81–83]. The active ingredients of a detergent are known as 'surfactants'. Builders are as equally important as surfactants. Their main function is to form metal complexes with divalent calcium and magnesium ions making them less available, and thus not interfere with the surfactant action [84]. The other ingredients include fluorescent whitening agents, enzymes and antiredeposition agents [85–87]. The components and functions of a typical synthetic detergent are described in Table 2.2.

Surfactant is a general term for substances such as soluble detergents in a liquid medium, dispersing agents, emulsifying agents, foaming

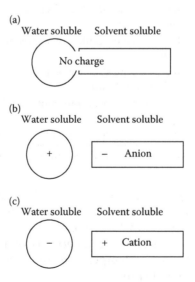

Figure 2.2 Three types of detergent molecules: (a) nonionic, (b) anionic and (c) cationic.

Table 2.2 Components and functions of a synthetic detergent

Components	Percentage (%) (approx.)	Functions
Surfactant	5–15	Loosen and disperse the soil, provide or control suds; basic 'cleaning' ingredients.
Builder	15–30	Soften water, aid surfactant in the dispersion of soil, buffer the detergent solution in the alkaline region.
Fluorescent whitening agents	<5	Overcome yellowness on the fabric and provide whiteness of blue-white hue.
Enzymes	Added as per the requirement	Catalyse the breakdown of protein- or carbohydrate-based stains to facilitate the removal by the surfactant and builder.
Antiredeposition agents	Up to 50	Prevent removed soil from redepositing on the fabric.
Sodium silicate, sodium sulphate, water		Carries free-flowing powder.

Table 2.3 Functions of various components of a surfactant

Name	Function
Dispersing agent	Increases the stability of a suspension of particles in a liquid medium.
Emulsifying agent	Increases the stability of a dispersion of one liquid in another.
Foaming agent	Increases the stability of a suspension of gas bubbles in a liquid medium.
Penetrating agent	Increases the penetration of a liquid medium into a porous material.
Wetting agent	Increases the spreading of a liquid medium on a surface.

Source: Lyle, D. S., *Performance of Textiles*, John Wiley and Sons, New York, 1977.

agents, penetrating agents and wetting agents [79]. All these ingredients are necessary in a good detergent. The functions of various ingredients in a surfactant are described in Table 2.3. The laundering process largely depends on the nature of the soiling material as well as the nature of the surfactant. Both alkalinity and temperature are important factors in conserving colours of the textile products.

Carbonates, phosphates, zeolites (aluminosilicates) and citrates are the most common builders used in laundry detergents [89,90]. Although the effectiveness of carbonates is debated, they are environmentally safe [91]. A drawback of carbonates is that they do not suspend soils in solution. Hence, the soils can be redeposited on the fabric surface leading to harshness and greying [92]. Phosphates are extremely effective but are banned in many countries as they are supposed to cause eutrophication of water systems. Zeolites and citrates are becoming important in recent detergents [93]. Zeolites can be used as a substitute or partial replacement for phosphates in areas where phosphates are banned. Citrates are becoming increasingly more important in liquid detergents [94].

Generally, laundry detergents come in two forms such as heavy-duty and light-duty. Heavy-duty (all-purpose) products are used for general laundry, whereas light-duty products are designed for hand washing delicate items or lightly soiled clothing [95]. In addition to soaps and detergents, alkaline salts or combinations of these are used for cleaning. These cleaning agents remove soil by one of the following mechanisms:

- Lowering surface and interfacial tension,
- Solubilisation of soil,
- Suspension and/or emulsification of removed soil, and
- Inactivation of water hardness, and
- Neutralisation of acid soil, particularly the saponification of fatty acids (soap formation).

Low-temperature and warm-water washing is a common practice nowadays compared to hot-water washing that was used previously [96]. The manufacturers of detergents suggest their use at a range of temperatures. However, specific soaps and detergents are more effective within a narrow range of temperature.

The effectiveness of a washing process depends on a range of variables related to the washing machine and nature of the detergent. The range of variables involved during washing includes: (1) washing conditions – washing time, temperature, rinsing time, agitation, temperature and volume of water; and (2) detergent type – nature of detergent, concentration, electrolyte and redeposition inhibitor [97].

Bruce et al. [98] evaluated the effectiveness of liquid detergents of varying formulations (unbuilt and built) in cleaning a standard soiled fabric in soft water (5 ppm). It was found that under the standard conditions, although not significant, the unbuilt liquid detergents produced better results in both warm and hot water. While washing in soft water, liquid detergents are not as temperature-dependent as powdered detergents in their ability to clean the samples.

The detergents used today have undergone several changes. Manufacturers are attempting to make sustainable products wholly or in part of natural raw materials such as oils from plant sources (i.e., palm, corn) or alcohols [99–101]. These new detergents perform effectively in cold water and produce better results compared to earlier detergents. These new detergents consists of builders, bleach activators and enzymes that work well at low temperatures. Table 2.4 compares traditional detergents with the contemporary detergent formulations.

2.1.4   Other washing aids

Laundering aids include bleaches, disinfectants, softeners and starch [2]. Bleaches are used to aid detergents in producing a cleaner and brighter fabric appearance [102]. The laundering bleaches most commonly used are liquid chlorine (strong), powdered chlorine (mild) and oxygen (weak) [102–104]. Chlorine bleaches can help to improve the whiteness of white clothing if used in a washing cycle [105]. However, they should not be used for coloured items. The hypochlorite ion, which is an oxidising agent, is the chemically active ingredient in liquid chlorine bleach [106]. These bleaches are the most effective stain removers but present a higher risk of damaging clothes. They should not be used for wool, silk, spandex, acetate fibres or their blends. Bleaches should be used in accordance with the care instructions to prevent damage to fabric dyes and finishes.

The action of powdered chlorine bleach is similar to that of liquid, but is gentler in action. The group of chemicals in these bleaches are N-chloro compounds, which release bleaching ingredients more slowly. Powdered chlorine bleaches should never be sprinkled directly on clothes as they can cause damage. Oxygen bleaches are the safest and maintain the brightness of white and coloured items if used regularly.

Disinfectants are used to reduce the amount of bacteria surviving in hot water and laundry detergents [107,108]. Bacteria may be spread by laundering if they are not killed by home laundry methods [97]. Quaternary ammonium, phenolic and pine oil disinfectants are commonly used laundry disinfectants [109]. Quaternary ammonium disinfectants should be added to the rinse water and phenolic disinfectants can be added either to the rinse or wash cycles. Pine oil disinfectants and chlorine bleaches should be added to the wash cycle. Liquid chlorine bleaches can also act as a disinfectant for fibres on which chlorine can be used. A very small amount of hypochlorite (about 20 ppm) in the washing liquid can destroy the bacteria present in a normal family wash load.

Softeners are used to reduce or eliminate the static charge in synthetic fibres, to make fabric softer, fluffier, and easier to iron, to minimise wrinkling and to help in preventing lint sticking to garments [3,110]. Softeners are also used to retain or improve the softness of the clothing such as towels, clothes with a velvet effect. A lubricating film is created on the fibres by the application of softeners, which allows them to move more readily against each other, thus making the fabric softer and fluffier [111–113]. In synthetic

Table 2.4 Comparison of detergents – traditional and contemporary alternatives

Component	Traditional	Contemporary
Surfactants	Anionic: soap, fatty alcohol sulfates, linear alkyl benzene sulfonate Non-ionic: alcohol ethoxylates, alkylphenol ethoxylates; e.g., nonylphenol ethoxylates (NPE), ethylene oxide/propylene oxide, alkylpolyethers	Anionic: Methyl ester sulfonate (MES) Nonionic: methyl ester ethoxylates alkyl polyethoxide (APE) Blends: Alcohol ethoxylates/ethanol amines
Builders	Tripolyphosphates, carbonates, silicates, zeolites, citrates, organophosphonates	Polycarboxylates, soda ash/silicate, polyacrylates, silicate/carbonate cogranule, hydroxyethyliminodiacetic acid (CEN 08)
Foam depressants	—	Silicones, soap for high-efficiency (HE) detergents
Anti-redeposition agents	Carboxymethyl cellulose	Acrylate polymers, carboxymethylinulin from chicory root, amylases for startch-based stains, cellulases for cotton fuzz removal, proteases to break down protein, stains, lipases for oil-based soils, mannases for carbohydrates of the mannan family (e.g., guar gum)
Bleach	Perborates, percarbonates + activator	Perborate+cold-water activator tetraacetyl ethylene, percarbonate with manganese-based activator
Colour protection	Polyvinylpyrrolidone	Dye transfer inhibitor-imidazole derivatives
Optical brighteners	Stilbene, azole, coumarin, pyrazoline derivatives	0.05–0.3% fluorescent shitening agents (FWA)in most commercial detergents
Soil-release agents	—	Soil-release polymers (SRP), delivered by the detergent coat low polarity fibre surfaces with a very thin layer of amphiphilic (hydrophilic) polymer
Filler	Sodium sulfate	Reduced amount of filler
Miscellaneous	Fragrance, colour beads, opacifiers, anti-caking agents to improve flow of powders	Many are fragrance free and/or have no colour beads

fibres, the lubricating film absorbs moisture from the air, which helps to reduce the generation of static. Fabric softeners can be added to the wash, rinse or drying cycle. Some softeners are manufactured to be compatible with detergents and other laundering aids. Softeners should be added in the approved concentrations and at the appropriate time of the cycle.

One of the oldest laundering aids is starch, which is still used in home laundering [114,115]. Starch is used to: (1) obtain a crisp, stiff and shiny fabric appearance, (2) help to keep a garment clean for a longer time, (3) replace the original finish applied to the fabric by the manufacturer and (4) facilitate stain removal as soiling is removed with the starch during washing. Starches may be classified as precooked vegetable starches, starch substitutes and aerosol starches [116].

2.2 Dry cleaning

Dry cleaning is the process of cleaning clothing items and other textiles using a chemical solvent other than water [117,118]. Dry cleaning is used to remove soil and stains from delicate fabrics, which cannot withstand the conditions used in the washing machine and dryer. As the name indicates, dry cleaning is not completely dry, rather it is performed with the use of various solvents and/or other chemicals instead of water.

In the majority of the cases, the solvent used is perchloroethylene or tetrachloroethylene. Perchloroethylene, or simply the 'perc' or 'PCE', has excellent cleaning abilities, is non-flammable, gentle to most garments and stable at the dry cleaning conditions [119–122]. However, perc was the first chemical to be classified as a carcinogen by the Consumer Product Safety Commission (a classification later withdrawn) [123]. In the year 1993, the California Air Resources Board (CARB) adopted regulations to reduce the emissions of perc from dry cleaning operations. The U.S. Environmental Protection Agency (EPA) also followed suit in the same year. The EPA updated this regulation in 2006 to reflect the availability of improved emission controls.

If a care label in the clothing indicates 'Dry clean only', it should be taken to the commercial dry cleaner to avoid any physical damage and for the best results. Otherwise, the clothing can be hand washed. However, the clothing should not be machine washed, which may result in shrinkage, colour loss, other damage and/or the fabric may lose its softness. Dry cleaning cannot remove all the stains and soiling from the garment. In some cases, the stains have the tendency to be permanently set in the fibre and the fabric, or the buttons and decorative beads can be permanently damaged due to dry cleaning. Hence, it is essential that consumers as well as the dry cleaners understand the care instructions before the dry cleaning process. The process has to be performed in accordance with the care instructions indicated by the textile manufacturers on the products.

A combination of lye, water, ammonia and a kind of clay was initially used to remove oil stains on the garments. All discussions on the origin of dry cleaning agree that it was the surprising removal of stains after pouring and the evaporation of a petrol-based liquid on a greasy fabric. Thus, 'dry cleaning' is a cleaning process that uses solvents to remove soils and stains on the articles. The misnomer arises from the avoidance of water for cleaning.

Textile products and garments received at the dry cleaning store should first be categorised by their colour. Light- and dark-coloured items should never be cleaned together as mentioned in the laundering section. Light colours should be cleaned using fresh chemicals to avoid colour weakness. Dark colours can be cleaned by previously used chemicals that have been thoroughly distilled and filtered.

The products received for dry cleaning are kept subject to the cleaning operation according to the symbols on the care labels and grouping. Owners of the product should be informed if the product is, or suspected to be, not suitable for dry cleaning operations, and the cleaning method of the product should be determined according to this notification.

2.2.1 Dry cleaning solvents

Fluids other than water are used in the dry cleaning process. In the early days, garment scourers and dryers identified several fluids that could be used as dry cleaning solvents. These included camphene, benzene, kerosene and gasoline, all of which are dangerously flammable [124–127]. Hence, dry cleaning was a hazardous business until safer solvents were developed. In the 1930s, perc (a non-flammable, synthetic solvent) was introduced and is still used today in many dry cleaning plants.

Perc has been widely used in dry cleaning since the 1940s and until today, it is the most common solvent. Other cleaning solvents such as hydrocarbons, modified hydrocarbon blends, glycol ethers, liquid silicone, liquid carbon dioxide (CO_2), brominated solvents and siloxanes are also being used for dry cleaning, which are considered as eco-friendly or green solvents [128–131]. In addition to these solvents, some other solvents used include camphor oil, turpentine spirits, benzene, kerosene, white gasoline, petroleum solvents (primarily petroleum naphtha blends), chloroform, carbon tetrachloride, trichloroethylene, 1,1,2-trichlorotrifluoroethane and 1,1,1-trichloroethane. However, these solvents are not in commercial use due to the environmental impacts, cost and other factors.

Generally, the dry cleaning solvents are used at the ambient temperature (about 20°C). Unlike the laundering, the solvent used in dry cleaning is not heated during the cleaning cycle. The heat is used only in the drying process to remove the excessive solvent from the garments. Later on, these solvents can be recycled by appropriate processes for reuse in the subsequent dry cleaning cycles.

2.2.2   Other chemicals

Other chemicals used during dry cleaning include detergents, chemicals for size retention and other speciality chemicals. The detergents perform the following functions during dry cleaning:

1. Help in the removal of water-soluble soils as they carry moisture,
2. Help in the suspension of soil after the soil is being removed from the fabric, and
3. Act as a spotting agent to penetrate into the fabric so that the stains are easily removed.

The detergents can be introduced into the dry cleaning machines by two different systems, namely: the charged systems and injection systems [132]. In the charged systems, the detergent is added to the solvent or 'charged' at a certain percentage on the weight of the solvent (1–2%). These systems use anionic detergents. In commercial operations, solvents containing anionic detergent or pre-charged solvents are being used. In injection systems, the solvent is added to the wheel of the dry cleaning machine saturating the garments, and the detergent is injected into the flow line or into the drum of the dry cleaning machine by a pump. Cationic detergents are appropriate for these systems.

Sizing chemicals are used during dry cleaning for retaining the size, shape, body and texture of the fabric. These chemicals are based on hydrocarbon resins such as alpha methyl styrene and styrene. The chemicals can be used either in the solid (powder) form or in the liquid form. The solid form is used with perc dry cleaning. The majority of the liquid sizing chemicals has a petroleum-solvent carrier even up to a volume percentage of 50. The sizing chemicals can be added in three different ways to a dry cleaning machine such as by a continuous bath in the machine, by dipping clothes in a tank of sizing chemicals and by spraying the aerosol form of the sizing chemicals on the garments after they have been dry cleaned. In the continuous bath process, 0.5–1.5 wt% sizing chemical is added to the dry cleaning machine, whereas, in the dipping process, 1–4 wt% is being used.

In addition, the liquid sizing chemicals carry optical brightening agents (OBAs) and antistatic agents [133]. OBAs have been widely used in laundry detergents for several years. Recently, they have been used in dry cleaning as well. The OBAs are used to improve the whiteness of the fabric after dry cleaning. The OBAs are normally added to the dry cleaning detergents or the sizing chemicals. Some other speciality chemicals such as bactericides and fabric conditioners are also used during dry cleaning to achieve some specific functions. In addition, antistatic and antilint agents (to prevent the build-up and retention of lint) are being used during dry cleaning. Some of the antistatic agents are based on the chemicals such as sulphonated polystyrene or sulphonated polystyrene/ maleic anhydride polymers. In some instances, fabric conditioners are being used during the dry cleaning to condition or restore the lustre and shine of clothing made of leather, suede and silk. These conditioners are based on petroleum naphtha or a perc-based solvent.

2.3 Dry cleaning versus wet cleaning

There is a fundamental difference between the wet cleaning and dry cleaning operations, which the consumers should thoroughly understand. The garment should be treated by a professional organisation (wet/ dry cleaning if they are labelled so). For several years the dry cleaners have wet cleaned a small percentage of their wash load either by hand or small washers. However, dry cleaning and wet cleaning differ in the use of solvents and other washing aids. In addition, they also differ in the degree of cleaning and effectiveness in removing special stains [134]. Each process has its own benefits and limitations as described below.

2.3.1 Benefits of wet cleaning

- No hazardous chemicals are used, hence no air pollution and lower water pollution [135].
- Easy removal of water-based stains; whites look whiter and better soil removal from some garments.
- A wide range of fibre types such as wool, silk, linen, cotton, leather/ suede, wedding gowns and garments decorated with beads can be wet cleaned.
- Wet cleaning much cheaper compared to dry cleaning as solvents are used in the latter process, which are more expensive than water.
- Wet cleaned clothes are free from chemical odor unlike dry cleaned clothes, which can retain strong odor after dry cleaning. Softening agents used give a pleasant smell to the clothes.
- Wet cleaning consumes about 50% less energy than dry cleaning, hence, is energy efficient.

2.3.2 Limitations of wet cleaning

In spite of these advantages, the wet cleaning suffers from certain disadvantages such as [1,17,136,137]:

- Shrinkage, wrinkling, surface changes, felting, loss of lustre and dye-bleeding problems can occur, which need special care to avoid the problems.

- The clothes may undergo change in the dimensions leading to shrinkage or stretching and improper fit.
- Although there are no organic solvents used, a large amount of water is used in wet cleaning leading to large quantities of contaminated wastewater.
- Unable to remove some hard grease, oils and wax-based stains.
- Can cause additional ergonomic risks to workers as it is labor intensive.

2.3.3   Benefits of dry cleaning

- Better cleaning efficiency compared to wet cleaning to effectively remove some stains, oil marks and greases.
- Reduces shrinkage, wrinkling, colour fading and distortion of the fabric.
- The original properties of the material are better retained.
- Protects texture and increases the durability of clothes.
- Dry cleaning facilities have started to use eco-friendly solvents and chemicals that disintegrate easily and are less harmful. These chemicals do not produces odor, or produce reduced odor, and clothes smell fresher and feel better.

2.3.4   Limitations of dry cleaning

The dry cleaning process has the following limitations [138,139]:

- People with sensitive skin may have negative reactions to the chemicals used in the dry cleaning process.
- The use of perc can enter the body through dermal and respiratory exposure leading to irritations of the eye, nose and throat; damage to the liver and kidneys; impaired memory; confusion; dizziness; headache; and drowsiness. Repeated dermal exposure can lead to dermatitis [140–142].
- Only skilled people can perform this as a higher health risk is associated with dry cleaning.

2.4   Drying

The process of removing excess water from clothes after washing or the final rinse is known as drying. Generally, the dry fabric is not free from moisture, but is in a state of dynamic equilibrium with the ambient atmosphere. The dynamic equilibrium represents a state of constant moisture content where the rate of moisture lost equals the rate of moisture absorbed. While drying the clothes, it is important to read the care labels and understand the conditions the fabric can withstand while drying. Inappropriate conditions can lead to the shrinkage of clothes or damage to the fabric. The clothes can be dried in shade or sunlight or by using a dryer. Drying the clothes in a dryer saves time and is essential when the climatic conditions do not allow the clothes to be dried outside. The following instructions can help to dry the clothes properly during drying. Various types of drying are discussed below [143,144].

Line drying: This is the simplest method of drying and requires the garment to be hung on a clothesline so that it can attain equilibrium with the ambient atmosphere. This is generally used in geographical areas where the ambient temperature is high enough so that the drying occurs in a reasonable time period. The time also depends on the amount of moisture to be removed, the material type, the wind speed and the relative humidity [111].

Spin drying or hydroextraction: This is done by subjecting the wet clothes to a centrifugal force [145,146]. The major concern in spin drying is that the applied force can result in crease formation and in certain cases result in permanent creases. Some delicate fabrics like silk can be damaged by this process. Spin drying has been found to be more effective in removing water and produces more consistent results. However, heat-set and wrinkle-resist garments are best candidates for this treatment.

Tumble drying: The residual liquid water after the clothes are spun dried in a washer is generally removed by turning the water into steam by heating and then extracting the steam [147,148]. Heating for a specific amount of time can convert the liquid water into vapour and can make the clothes completely dry. This principle is used in drying the clothes using equipment known as a tumble dryer. While drying, a massive amount of hot, humid air is generated, which needs to be effectively extracted from the drying chamber [149]. A huge amount of electrical energy is needed to produce steam from water.

While using a tumble dryer, the lint trap (or the lint screen) should be always cleaned before the start of the drying cycle. This helps the dryer to work more efficiently, and any chance of fire is reduced as dryer lint is very combustible. While adding the load to a dryer, shaking out the clothes before putting them into the dryer helps to prevent wrinkles and reduces drying time. The dryer should never be overloaded in an idea of saving time as it will have the opposite effect [150]. The overloaded clothes will take longer to dry and get more wrinkled as enough room is not available in the dryer for the clothes to fluff out.

A dryer sheet can be added to the dryer, if liquid fabric softener is not used during the wash cycle. The dryer sheet helps to soften the clothes, as well as reduce the static cling. The correct cycle should be selected for the clothes depending on the fibre type such as: (1) air dry cycle: should be selected for fluffing pillows or refreshing clothes as the heating is the minimum; (2) gentle cycle: for delicate items like lingerie and workout clothes; (3) permanent press cycle: for synthetic fabrics; and (4) cotton cycle: for towels, jeans, sweats and other heavy fabrics as the heat is at the maximum in this cycle. When the drying cycle is completed, the clothes should be removed from the dryer as soon as possible to prevent wrinkles. The clothes should be neatly folded or hung to avoid wrinkles.

Outdoor drying: The earliest method of drying clothes, still used today, is outdoor drying [144]. Outdoor drying or air drying of clothes has several advantages such as: (1) no electrical energy is used, hence it is ecofriendly; (2) it generally leaves the clothes fresh; and (3) the house is free from being damp. The disadvantages of outdoor drying include: (1) a longer time taken to dry, which can range from a few hours to several hours or even a day or more; (2) the chances of rain water wetting the clothes; (3) the risk of theft; and (4) the possibility of air pollution making the clothes dirty again.

Clothing can be dried at any temperature above the freezing temperature, which is accentuated by dry air. The best conditions for outdoor drying include warm, windy conditions, when the humidity is relatively low. Hence, summer days are more efficient in drying the clothes than the winter days. Outdoor drying in winter may instead cool down the water and turn to ice, which in turn slows down the drying process. Hang drying the clothes helps the air to move faster around the clothes as compared to flat drying.

Indoor drying: The indoor drying of clothes is still widely used in many places. However, this method of drying has several disadvantages rather than advantages, which are discussed below:

- Drying clothes indoors takes more time than outdoor drying.
- Energy is needed to dry the clothes indoors, which should be derived from some sources. The energy needed to evaporate water from the clothes is derived from the ambient air, which results in cooling down slightly [151,152]. This in turn results in more cost of heating the room in cold.
- The water vapour that evaporates from the clothes condensates as damp or mould on the walls. This can be avoided by opening the windows, which in turn causes more expense because of the loss of heat energy.

People are encouraged to dry more clothes indoors during prolonged wet weather to reduce the fuel bills involved in using tumble dryers. This could pose health risks especially in people prone to asthma, by increasing moisture inside the living area that encourages moulds and dust mites. The indoor drying of clothes that contain fabric conditioner is likely to increase the amount of cancer-causing chemicals in the air.

Indoor drying can also lead to increased energy usage as radiators are often turned up to help the drying process, which in turn worsens the fuel consumption. It is a good practice to dry the laundry outdoors whenever possible, or to use energy-efficient, condensing tumble dryers when outdoor drying is not possible. If the clothes need to be dried indoors, they should be placed in ventilated areas where an abundance of natural light is available and, if possible, heat is also available.

2.5 Pressing

Garments are pressed to remove any creases and present the garment in an attractive condition suitable for sale. Garment presentation to the consumer is a vital step in the finishing of a product [153]. The opinion of the customer is an integral step in brand recognition. A poorly presented product will have a detrimental effect on the brand's quality and therefore product saleability. A badly creased garment will lower its retail value and thus the manufacturer's sale margin. Pressing therefore is an important step in the production process. Pressing should accomplish the following:

- Removal of all manufacturing creases and wrinkles.
- Clarity of pleats if there are pleats present (such as in skirts and trousers).
- Uniformity of collars and cuffs if present.
- Stabilising the garment, particularly in the case of wool knitwear, to retain the desired shape.
- Relaxation of any stresses induced during the garment manufacture.

In order to achieve good pressing quality, there are four basic parameters that need to be controlled to meet optimum performance, which include heat, moisture, pressure and finally cooling with a vacuum. The importance of each parameter is discussed as below.

Heat is required in most pressing operations to enable the fibres to soften and thus stabilise the garment shape. Temperature selection is of utmost importance as an incorrect temperature setting can cause damage to fibres and yarns.

Moisture is introduced by the use of steam. Steam at different pressures has different moisture contents. The higher the steam pressure, the lower the moisture in the steam. The presence of moisture is required to aid in fibre swelling and thus shape stabilisation [154]. Different fibres require different amounts of moisture. For example, natural fibres, such as cotton and wool, and regenerated cellulose fibres, such as bamboo viscose and viscose rayon, require the presence of moisture in the steam and therefore, steaming tables are usually preferred. On the other hand, synthetic fibres require heat to promote swelling and therefore relaxation of the structure. Excessive moisture may cause fabric shrinkage and colour bleeding.

Pressure is applied to the garment during pressing to give good crease retention and permanency [154,155]. Excessive pressure may result in garment or crease distortion.

Vacuum is applied at the completion of the pressing operation. This draws cool air through the garment, reducing the garment temperature, lowering the moisture content and increases shape retention. Particularly for garments made from wool and wool blends, this also applies to cotton and viscose blends with synthetic fibres such as polyester and nylon.

2.6   *Cleaning of protective textiles*

The technological developments have helped to achieve improved performance of the technical textiles by the use of advanced materials, finishes and techniques. However, the method of care and maintenance of these protective textiles, which greatly influences the protective performance, has not changed much. The severity of the laundering processes, washing parameters and the chemicals used should be carefully decided to retain the properties. In many cases, the use of high wash temperatures, excessive detergent, chlorine bleach and other attempts to remove soil, stains and odours can deteriorate fabric properties.

Appropriate care and maintenance of personal protective clothing (PPC) is important to the manufacturer, supplier and critical to the user. The safety of the user depends on the performance of the PPC to a great extent. In addition to proper cleaning, the care and maintenance procedure should remove the contaminants from the PPC that may affect the performance. This is essential to maintain the performance throughout the life of the PPC. Hence, to achieve this, the selection process for any PPC should consider the nature of contaminants and soiling, including their care and maintenance procedure.

Depending on the nature of the protective textile, the method of cleaning varies. The protective textiles might include a care label describing the methods of cleaning and things to be avoided [156]. Following the care instructions will help in retaining the properties of the protective textiles. Commercial detergents contain brighteners, fragrances, softeners and other additives, which may leave residues, leading to clogged pores in the fabric. Hence, specially formulated detergents that help to maintain the performance should be used.

Some of the protective textiles are coated with specialty coatings to improve the performance and/or comfort [157]. These items can be cleaned by general laundering or dry cleaning as per the instructions on the care label. However, in many cases, special care is needed during these processes as the coating is sensitive to temperature, chemicals and mechanical action. Furthermore, ironing is not possible in these coated materials, as it may negatively affect the coating. The traditional detergents used for apparel fabrics contain softeners, fragrances and brighteners that may not be appropriate for PPCs as they can leave residues, leading to clogged fabric pores. The washing chemicals and other aids that may otherwise deteriorate the properties of the finishing should be properly removed from the clothing. The detergents specially formulated or recommended for the PPCs should be used so that functionality is maintained.

The presence of sweat, food, chemicals and blood stains is obvious due to the protective function of the PPCs. Any attempts to remove these by laundering or dry cleaning using inappropriate chemicals or methods can ruin the performance. The other problem associated with cleaning of the PPCs is the unknown chemical formulation. In several cases, the composition of the detergent or solvent and additives are not known to the user, which can cause serious concern about the clothing.

Another important consideration while cleaning the protective clothing is retaining the comfort properties in addition to the performance [158]. The cleaning methods adopted can change the dimensions, shape, physical properties or the texture of the fabrics, which may impair the comfort features of the clothing. The protective clothing may become uncomfortable to wear for extended periods, which may lead to the rejection of the same during the intended work. This increases the risk of hazards during the regular work. Hence, the personal protective clothing should be cleaned in such a way that it retains its: (1) protection level, (2) shape and dimensions and (3) comfort features.

The application of various functional finishes to protective textiles can alter their wicking behaviour and wettability, which in turn can affect the launderability of these items. The changes in the surface properties and pore volume are the major causes of the change in the properties. The nature and concentration of the finishes applied to the fabric affects the degree of change. For example, Rhee et al. [159] investigated the effect of a durable-press finish, stain-repellent finish and an antistatic finish on the laundering performance.

Apparel textiles are often rejected due to their change in appearance during wear, cleaning and storage [159,160]. However, for protective clothing, more than appearance, protection is the main requirement. The rejection of these garments is based on the amount of loss of the protection level or failure to meet the standard specification. Improper cleaning methods may not properly clean the items, which may affect the performance. For example, residual oil in a PPC worn by an oilfield worker may make the PPC flammable and fail to meet the specifications.

In several instances, the PPCs are rejected on the basis of general observation rather than performance evaluation as performance evaluation involves destructive objective tests. Hence, subjective observations by experts decide the future fate of the PPC, which depends on the interpretation of the change in appearance. Therefore, a PPC approved as usable by the expert may fail to meet the specification or an unapproved PPC may be suitable for usage. For the protective clothing with multiple layers such as firefighter's PPC, or PPC for cold weather protection, this subjective method is problematic as the observation of the internal layer(s) is very difficult.

The PPCs should be regularly cleaned before they are very dirty or heavily contaminated, otherwise it will be very hard to clean them. If embedded particulate matter or oily stains are not promptly removed, oxidation of the oils will make them less soluble and very hard to remove [161]. The protective clothing with oil stains may not fulfill the flame resistance [162]. Almost all the protective clothing should be regularly cleaned and stored in a ventilated area away from heat. They should never be stored without cleaning.

The PPCs can be cleaned by domestic or industrial laundering processes. For achieving the best results, the parameters used for laundering and drying should be established experimentally. These parameters are essential for certification of the protective clothing. The establishment of a care and maintenance procedure for various PPCs is a tedious process. The steps involved are: (1) analysing the requirements of the PPC, (2) identification and selection of the right PPC to provide the required protection and (3) the establishment of the right protocol for the care and maintenance of the care procedure. The first two steps are accomplished by reviewing the standards, Internet searches, legal requirements, peers in the groups and industrial practices based on the nature of the hazard. For the final step, it is essential to have the knowledge of the fibres, fabrics, other materials, garments and finishes used to prepare the PPC.

Many organisations dealing with PPC consider care and maintenance procedure as one of the key criteria when purchasing a PPC. It is essential to understand the types of soils that will be coming in contact with the PPC during the work and the cleaning process for the same. It is always imperative to select industrial laundering for the PPC as these facilities use detergents and equipment that are not used in home laundering. In addition, industrial laundering facilities provide repair and inspection services that can cover repairing of minor damages such as tears, holes and cuts. These processes can increase the serviceability of the PPC.

Industrial laundering facilities can deal with almost any types of stains and soiling deposited in the PPC. It is essential for these facilities to understand the type of the soil in the PPC for effective removal, to avoid any possible damage to the fabric and health hazards to the personnel. These laundries should collect the care and maintenance instructions from the manufacturer in order to prevent any damage to the protection performance of the PPC. Furthermore, they can also assist in the testing and evaluation of the PPC to monitor the performance over their life cycle.

In several instances it may be possible that the grease or oil stains or other contaminants are not sufficiently removed from the PPC. Hence, commercial dry cleaning facilities should be used for the effective removal.

If home laundering by the employees is selected for care and maintenance, it is essential for the employees to completely understand the process and parameters for the same. Initially, discussions should be organised with the manufacturer or supplier for a thorough understanding of the care instructions. In addition to the care instructions on the PPC, the employees should be provided with written instructions. This will help other people in cleaning the PPC instead of the employee. The restrictions on the PPC, such as to avoid bleach or high heat, should be fully understood by them. For any hard to remove soil or stains, the special process to be followed should be explained to the employees.

The number of laundering cycles has a direct impact on the performance of the PPC. Hence, the launderer should keep a record of the number of cycles. Some of the manufacturers provide a guarantee of the serviceability to a specific number of washing cycles. Once these numbers of cycles are reached, the PPC should be removed from service as they may not be able to render the desired protection. However, suitable inspection methods should be established to find out the end of life of PPCs.

The users of PPC can check the general features such as the working of zippers, snap fasteners, buttons, opening of seams, presence of cuts or tears and the presence of stains or other chemicals (from odour). This self-inspection can be performed on a regular basis such as weekly or monthly, or the PPC can be inspected by the qualified people in the organisation. If during an inspection it is found that repairing is needed, it should be done using the right material and method so that the performance is not deteriorated. It is a good habit to keep the details of the repairs of each PPC so that the performance can be detected throughout its lifetime.

Each organisation should establish their care and maintenance protocols for PPC. They should also keep a record of the new, in service and used PPCs. It is essential that the PPCs returned for storage should be cleaned and the storage area must be dry, clean and away from direct sunlight.

In many instances there may be remaining marks of stains still present on the PPC or a discolouration of the PPC. This does not indicate the PPC has met its end of life. A proper quantitative evaluation is essential for the same. The end of life of a PPC can also be set by the organisation by fixing a retirement date from the service date or the conditions the garment needs to be removed from service. After removal from service, the PPCs should be disposed or destroyed in an appropriate manner to avoid disposal and sensitivity issues.

There are several standards that cover the care and maintenance procedures used for different types of PPC. Some of the standards cannot be used as standalone for the PPC. They have to be used in combination with other standards. For example, EN 340 used for the ageing and other selection criteria of PPC is used in combination with International Organization for Standardization (ISO) 11612 for heat and flame protection. The following section describes the care and maintenance procedures to be followed for PPCs used for ballistic, stab, chemical and antibacterial protection.

2.6.1   Cleaning of firefighter's clothing

The useful life of a firefighter's personal protective clothing (FFPPC) is the time for which the FFPPC provides acceptable protection. Factors such as material, design, degree of exposure and intensity of flame, the maintenance and storage procedures affect the useful life of FFPPC. The frequent use of the FFPPC makes it soiled with contaminants and body excretions. This in turn can reduce the protection needed leading to flame resistance failure. Hence, proper care and maintenance is needed for FFPPC to remove the contaminants. This section will deal with the care and maintenance of the FFPPC and its influence on its performance.

The FFPPC may consist of multiple layers with different functions. The washing protocols used for one layer may not be appropriate for the other layers. In some instances these layers are detachable, and can be washed separately. However, in an integrated FFPPC with multiple layers, appropriate protocols are necessary.

The firefighter's clothing may be either home or industrial laundered to successfully remove most types of flammable and non-flammable soils [163]. However, home laundry detergents may not always be successful in removing some types of soiling such as heavy greases and oily soils. If flammable soils are not completely removed, the flame resistance of the garment may be compromised. The thermal protection of the flame-resistant clothing can be compromised by the contaminants on the surface. Although the fabrics may satisfy the requirements of flame protection as evaluated by the standard test, the presence of flammable contaminants in the uniform can burn easily, until consumed, leading to inadequate protection to the wearer. Hence, it is necessary to clean these flammable contaminants by appropriate methods prior to reuse. Laundering and scrubbing are the major maintenance procedures for firefighter's protective clothing. It is to be emphasised that laundering affects the ageing process chemically and mechanically to a significant level.

The fabrics selected for the preparation of the FFPPC should be able to withstand the protocols used for washing and dry cleaning [158,163]. In addition, the finishes applied should also withstand these processes. Washing detergent, the load, temperature and rinsing cycle should be carefully selected. The use of softeners, starch, bleach and other washing aids should be carefully considered as they can alter the performance of flame protection. Softeners and starches can deposit on the surface of the fabric altering the performance, whereas exposure to bleach can destroy the luminescence of the surface. Heavily soiled ensembles with particulate or abrasive soils should be pre-washed at 40°C initially, which will help to reduce abrasion in the wash wheel.

There are a gamut of detergents designed to be used in a range of washing temperatures with no adverse effect on the FFPPC. The detergents and washing protocols should be carefully selected to clean the soiled garment thoroughly, even considering a supplemental alkalinity and higher wash temperature. Light and dark colours in the load should be isolated as well as heavy- and light-soiled items. Loading the washer lower than the maximum capacity will provide the best results. The wash temperature is selected depending on the degree of soiling and the nature of the material used in the FFPPC. Although, a higher temperature is better for heavily soiled FFPPC, the care instructions and the compatibility with the washing chemical should be checked. In some instances, a higher temperature may cause problems of shrinkage, colour loss or change in the appearance.

A majority of the detergents commercially available are alkaline with a pH value ranging from 9 to 13, which effectively remove the soil and other contaminants in the fabric. A higher temperature, longer cycle and supplemental alkalinity can be used to remove more aggressive soil and oil marks from the FFPPC. The use of soft water is recommended for FFPPC [52,164]. The use of soaps can form insoluble scrums with hard water, which gets deposited on the fabric. These scrums can adversely affect the flame protection of the flame retardant (FR) clothing, as they are flammable. The protective clothing should be washed and dried inside out, which will help in retaining the properties of the outer layer. While laundering, the wash formulas and load sizes should be established to minimise redeposition and fabric abrasion. Short extract time can prevent wrinkle formation.

Residual alkalinity in the garments can cause skin irritation and other problems [53]. Hence, it should be neutralised by the use of sour in the final rinse cycle. The washed load should be thoroughly rinsed to remove the washing chemicals. The combination of thorough souring and good rinsing will reduce the chances of dermatological reactions from higher pH. In addition, if there is chemical residue in the FFPPC, the FFPPC can be tested for FR as mentioned in the American Society for Testing and Materials (ASTM) D6413, the standard test method for flame resistance of textiles (vertical test). This test is performed in an enclosed cabinet, where a controlled flame is exposed to the cut edge of fabrics for 12 seconds. Five specimens are selected for each length and width direction, which are mounted in metallic frames covered in three sides.

Parameters such as char length, afterglow and after flame are measured to describe the FR performance.

ASTM 6413 is helpful to find out the performance of a single FFPPC or even a group of aftercare producers. However, if the service history of the group differs, each garment should be tested separately.

The tumble drying conditions should also be carefully selected. Overdrying in many instances leads to shrinkage and hence should be avoided. Overloading of the tumble dryer can reduce the drying efficiency and result in improper drying. The removal of the FFPPCs while they are slightly damp (5–10% moisture) and hang drying will produce good results. The dried load should be instantly removed after the cycle is completed. Similarly, the pressing conditions should also be carefully selected to retain the appearance and protection performance of the FFPPC. There are several standards for the care and maintenance of the firefighter's protective clothing followed around the globe as described in Table 2.5.

Table 2.5 Various standards for care and maintenance of firefighter's protective clothing [158]

Standard	Purpose
ASTM F1449	Standard guide for industrial laundering of flame, thermal, and arc-resistant clothing.
ASTM F2757	Standard guide for home laundering care and the maintenance of flame, thermal and arc-resistant clothing
CEN/TR 14560: 2003	Guidelines for selection, use, care and maintenance of protective clothing against heat and flame.
ISO/TR 2801: 2007 (AS/ NZS 2801: 2008)	Clothing for protection against heat and flame-General recommendations for selection, care and use of protective clothing.
NFPA 1851	Standard on selection, care, and maintenance of protective ensembles for structural fire-fighting and proximity fire-fighting
NFPA 1855	Standard for selection, care, and maintenance of protective ensembles for technical rescue incidents
NFPA 2112	Standard on flame-resistant garments for protection of industrial personnel against flash fire
NFPA 2113	Standard on selection, care, use, and maintenance of flame-resistant garments for protection of industrial personnel against short-duration thermal exposures.

ISO, International Standards Organisation; TR, Standardised document for information and transfer of knowledge; CEN, Comite Europeen de normalization (European Committee for Standardization); NFPA, National Fire Protection Association; AS/NZS, The Joint Australian/ New Zealand Standard.

ASTM F1449, developed in 1992 by the ASTM international subcommittee F23.80, was useful in providing care and maintenance information for industrial laundries. This standard describes garments soiled by flammable substances such as solids, solvents, oils and petrochemicals with a flammability risk. If residues of these chemicals remain, it can help in increasing the flammability risks. These chemicals have strong odours, which help to detect the presence of residual chemicals after laundering. Although this standard is helpful for industrial laundering, it provides limited information related to home laundering.

ASTM F2757, developed by subcommittee F23.80 in 2009, provides guidance for home laundering. This standard is intended for the use of those who choose to select a home laundering programme for flame, thermal and arc protective clothing. Although this standard does not recommend a home laundry procedure, it suggests following the garment manufacturer's care instructions. In addition, this standard lacks the information on other problems such as stain removal instructions; use of bleaches; use of soft water for laundering, etc. The standard along with the manufacturer's instructions can serve the purpose for home laundering of the FFPPC.

Several standards describe the test method to inspect the durability of the FR finish subjected to repeated laundering. For example, the American Association of Textile Chemists and Colorists (AATCC) 135 (dimensional changes of fabrics after home laundering) is the most widely used American standard. This standard is referred to in several other standards for FFPPC and other PPCs such as ASTM F1930, ASTM F1506, ASTM F2303 and the National Fire Protection Association (NFPA) 70E.

The protocol mentioned in AATCC 135 was unable to keep pace with the change in consumer choice. Hence, a set of guidelines was devised by the AATCC and published as a separate monograph in the AATCC technical manual. Monograph M6 from AATCC (the standardization of home laundry test conditions) provides a set of guidelines for laundering and drying for many types of garments or fabrics. ISO 6330 (textiles – domestic washing and drying procedures for textile testing) is the international standard similar to AATCC 135. This standard specifies protocols for fabrics, garments and other textiles that are home-laundered and dried. There are 10 and 11 different washing procedures for front-and top-loaders, respectively. In drying there are five processes ranging from line to tumble drying.

ISO 15797 specifies test protocols and equipment used for the evaluation of workwear of cotton and polyester/cotton (P/C) blends in industrial laundering. This standard can also be used for FR clothing prepared from the blends of other natural and synthetic fibre blends. NFPA 2112 describes that the FR garments should be tested for performance both before and after 100 cycles of washing as well as drying. The laundering protocol described in this standard is alkaline, which is used for heavily soiled PPCs (Table 2.6).

The FRPPC can be broadly classified into two groups, namely inherent FR or treated. The inherent FRPPC is prepared from synthetic fibres, which are synthesised with FR chemicals in their molecular structure during manufacturing. These materials are highly resistant to ignition or burning. On the other hand, the treated FRPPCs are prepared from natural fibres and treated with a FR finish in the fibre, fabric or garment stage. The examples of the two classes include clothing prepared from Nomex (inherently FR) and FR-treated cotton clothing such as Proban. Other synthetic fibres such as polybenzimidazole (PBI), Basophil, Kermel and carbon or oxidised polyacrylonitrile are inherently FR. In many cases, the natural fibres can be blended with inherently FR fibres to balance between the performance and comfort. Other fibres such as

Table 2.6 Specifications for industrial laundry formula specified in NFPA 2112 for testing the durability of FR textiles

Operation	Temperature (°C)	Time (min)	Water level	Quantity per wash load (g)
Break	66	10	Low	
Sodium metasilicate or equivalent				17
Sodium tripolyphosphate				11
Tergitol 15.S.9 or equivalent				22
Drain		1		
Carry-over	66	5	Low	
Drain		1		
Rinse	57	2	High	
Drain		1		
Rinse	48	2	High	
Drain		1		
Rinse	38	2	High	
Drain		1		
Sour	38	5	Low	
Sodium silicofluoride				6
Drain		1		
Extract		5		

Source: NFPA 2112-2012. (Standard on flame-resistant clothing for protection of industrial personnel against short-duration thermal exposures from fire)

polyester, viscose or modacrylic can also be treated with inherent FR fibres to improve comfort, durability and reduced cost. In some cases, all of these blending conditions may be selected for similar working conditions.

The inherent FRPPC are not much affected by the detergents and chemicals used during laundering or dry cleaning. But, the residual chemicals or unremoved contaminants on these items may reduce the flame protection. The treated FR fabrics gradually deteriorate if exposed to conditions such as the use of chlorine or peroxide bleach. Inherently, flame-resistant materials may change in colour or lose strength to some extent, but their FR properties are retained with these chemicals.

The cellulosic fabrics treated with FR, Proban, which are used for fire risks for a short duration, need special care to maintain flame retardant characteristics for 50 washes. However, several independent tests have shown that these garments pass flammability tests even after 100–150 washes. On the other hand, garments can fail in the flammability tests after just a few washes if they are not washed properly. The following care should be taken during laundering Proban garments:

- Proban-treated garments can be laundered in any conventional washing machine.
- Use cold or warm water for both whites and colours (temperature below 60°C).
- Select a wash cycle used for non-colourfast clothes.
- Only synthetic detergents (e.g., Dynamo, FAB, OMO, Radiant, Spree or Surf) should be used.
- For heavily soiled articles, a short (up to 2 hours) pre-soak cycle may be useful.
- Regular washing will prevent soiling buildup.
- Garments can be tumble dried (avoid over-drying as excessive shrinkage may occur).
- Garments may be dry cleaned.
- Do not wash garments in traditional soap-based powders (e.g., Lux, Velvet, Advance) as the soap powders can form flammable deposits that may adversely affect the flame-retardant performance of the fabric.
- Do not use hypochlorite bleaches as they can damage the finish and can lead to the flame retardancy becoming ineffective.

While home laundering, a normal or cotton cycle can be selected at any water temperature up to a maximum of 60°C using a typical home laundry detergent. The use of tallow soaps containing animal fats should be avoided. The garments should be turned inside out before washing to reduce the chance of abrasion. The use of starch or fabric softeners should be avoided as they may mask the performance and the particles may facilitate burning.

It is essential that all the contaminants are completely removed from the ensembles after the wash process. In some cases, the use of stain removers is essential for complete cleaning. In addition, the use of hot water can make detergents more effective in removing soils. If it is not possible to remove all contaminants in home care, the garments should be dry cleaned using perc. If garments become contaminated with flammable substances, they should be immediately removed and replaced with clean flame-resistant ensembles.

The presence of stains or odours after laundering indicates improper cleaning of the clothing. As the test involved in evaluating the performance is destructive, it is not possible to assess their performance after care procedures. Stains and contaminants remaining after laundering can lead to discolouration of the fabric. Hence, care should be taken to properly clean the items to avoid contaminants catching fire. Dry cleaning can be used to remove oils and greases, which are hard to remove. It is always better to consult the manufacturer of the ensembles for detailed instructions in the case of difficulty in wet or dry cleaning.

2.6.2   Cleaning of body armour

The ballistic panels of soft body armour become saturated with perspiration due to continual field use. The constituents of human perspiration are mainly water along with small quantities of organic compounds and inorganic salts. Human perspiration in the long-term can affect the properties of ballistic materials [165]. The laundering of body armour is not recommended by the manufacturers; even some standards specify not to clean them. For example, the National Institute of Justice (NIJ), who publishes standards for the testing of body armour, specifies not to clean the ballistic panels and coverings. However, the carriers can be washed and dried with conventional home-laundering techniques. Hence, after several times, the wearers can attempt to deodorise and clean the body armour by spraying the panels with odour neutraliser, cologne, disinfecting sprays and/or wipe them with dilute solutions of detergent or bleach.

Studies have shown that not only ageing reduces the ballistic performance, but also the method of care and maintenance affects it. The NIJ standard provides guidelines to evaluate the performance of body armours in use. The guidelines provided by NIJ standard, can help to extend the useful life of the armour that may save someone's valuable life one day. These suggestions are discussed below [166]:

Dos	Do nots
Drip dry the armour indoors.	Machine wash or dry the armour, as machine washing can alter the ballistic performance.
Regularly inspect the armour for cuts, tears and other damages to the carrier and ballistic elements.	Bleach the armour or use products containing bleach for care and maintenance.
Contact the manufacturer with any questions about care and maintenance of the armour.	Use commercial laundering facilities as they may use harsh chemicals that can affect the armour's protection performance.
Follow the manufacturer's instructions for the care and maintenance of the armour. A person should be aware of the cleaning methods before doing it.	Dry clean the armour as dry cleaning solvents can affect the armour's protection performance.
	Dry the armour outdoors as some ballistic fabrics degrade as a result of ultraviolet (UV) exposure.
When necessary, hand wash the armour with a mild detergent in cold or warm water. Rinse it thoroughly to remove all traces of detergent.	Do not attempt to repair the armour. Armour should be returned to the manufacturer for repairs or replacement.

When washing the whole body armour is allowed (such as combat clothing used for Tier 1 and UBACS (Under Body Armour Combat Shirt)), they are routinely laundered. It is important to closely follow the care instructions provided by the manufacturer as there is a strong impact of the care process and chemicals on the performance [167]. Appropriate methods of cleaning and storing can help to maximise its service life and effectiveness of protection. It is always very important that the users become better educated on how their vests should be cared, stored and maintained. It's an investment that can save an officer's life.

The laundering of body armours negatively affects the physical and mechanical properties like the apparel fabrics [168]. The effect of multiple washing and drying cycles on the fragment protection was analysed by Helliker et al. [169]. Woven para-aramid (a), para-aramid felt (b), hydro-entangled UHMWPE (ultra-high-molecular weight polyethylene) felt (c) and single jersey knit silk (d) was used in this study. It was observed that the ballistic performance (measured using 0.24 g of fragment simulating projectile (FSP) of fabric (c) and (d) was not affected by the laundering up to 27 laundering cycles. However, both the fabric (a) and (b) showed improved ballistic performance against 0.24 g FSP after nine laundering cycles. Furthermore, the reason for the increase in the performance was uncertain and it was assumed that the increase in friction between fibres/yarns due to the removal of the lubricants used in knitting/ weaving or due to the surface peeling of the fibres might have resulted in the increased value.

This study also established significant change in the physical properties due to laundering. Shrinkage was observed in all the fabrics, which lead to the increase in the thickness of the samples. The increase in the thickness did not change the ballistic performance of the fabrics. However, it might have changed the thermal resistance of the garments and resulted in change in the appearance. No significant changes in the mass of the fabrics were observed, which indicated no loss of fibre during laundering. Ballistic protective clothing comprised of felts (i.e. fabrics b and c) was more severely affected by laundering than the woven or knitted fabrics examined in the study [170].

Fibres such as silk and para-aramid are degraded by laundering due to the surface peeling of fibres, which may lead to the loss in tenacity of yarns and fabrics. The surface peeling of fibres may increase the friction between the yarns, which can increase the ballistic performance of the fabrics after laundering. Other possible causes of change in the ballistic performance can be attributed to the removal of the lubricants from the fabric that was used during weaving or knitting. It has been earlier shown that the scoured fabric has improved ballistic performance to identical fabric in the loom state due to the removal of the lubricants.

The body armour can be designed by placing the protective layer inside a washable and breathable carrier. The carrier is generally made from cotton or other absorbent material to provide wicking of the sweat. The carrier can help in improved breathability in addition to the ease of care and maintenance. The carrier can be separated from the armour panel for care and maintenance. However, the armour may become bulky without the significant contribution of protection by the carrier.

The effects of various cleaning chemicals such as odour neutraliser, liquid detergent and chlorine bleach; and perspiration on the tensile, chemical and surface morphological properties of different ballistic materials (aramid yarns and fabrics, UHMWPE yarn and piperonyl butoxide (PBO) yarn) were investigated by Chin et al. [165]. It was observed that the tensile properties of the aramid and PBO materials decreased after exposure to plain water, artificial perspiration, detergent, odour neutraliser and chlorine bleach. The tensile strength of UHMWPE was decreased only by chlorine bleach. Chlorine bleach resulted in a significant decrease in the tensile strength for all the materials compared to plain water. Chlorine bleach caused physical changes in the fibre surfaces, which may have caused the damage to the fibres.

Significant chemical changes initiated by strong oxidisers in the bleach were observed in the Fourier-transform infrared (FTIR) results. Aramids and PBO showed hydrolytic degradation, whereas UHMWPE showed oxidative degradation, which resulted in the change in fibre tensile properties. It was concluded that exposure to chlorine bleach over a period of time could significantly damage the soft body armour of aramid, UHMWPE and PBO fibres. Hence, chlorine bleach should be avoided in the routine care and cleaning of armour. Although additional damage was not observed on the ballistic fibres exposed to aqueous-based cleaning and artificial perspiration beyond that of water alone, it should still be noted that water can degrade the mechanical properties of aramid and PBO fibres long-term. Hence, the use of water and/or any aqueous-based products should be avoided or minimised in the care of soft body armour.

2.6.3 Cleaning of chemical-protective clothing

The amount of research on the cleaning of chemical-protective clothing is limited. The chemical-protective clothing can be contaminated during the application of the toxic chemicals, which may be absorbed by the fabrics. Wearing the contaminated clothing can result in higher levels of exposure to chemicals compared to clean clothing [164,171]. The amount of soiling depends on the chemical nature of the textile fibre, chemical treatments of the fabric and the type of chemical (i.e., whether oil- or water-based). The penetration of soil, entrapment of soil in the fibre structure, and/or in the spaces of fibres worn by mechanical wear during laundering or during use and chemical reaction of soil with fibre and finish can cause difficulty in soil removal.

The soiling of chemical-protective clothing can occur when the textile is worn during chemical applications or during laundering by cross-contamination. During laundering, the chemicals can be transferred from a soiled cloth via the washing solution to another. It can also occur by the redeposition of soil removed from the area of soiling into the washing medium, and thus to all areas of that cloth including the other clothes [172]. Some clothing retains the soil residue in laundered fabrics regardless of temperatures, detergent type, additives, pre-rinse or wash cycle, pre-wash treatment, fibre content, textile finish and the type of fabric.

In some instances, pre-rinsing can help in better removal of chemicals from the clothing. Pre-rinsing can involve an additional cycle in the washer, soaking in a container prior to the wash cycle or rinsing under a stream of running water. Several researches have shown that pre-rinsing helped in improving the cleaning efficiency of the chemical-contaminated clothing.

The concentration of the laundering chemicals affects the contaminant removal during laundering. It was observed that the degree of difficulty in the removal of the chemicals increases with the increase in the concentration of the chemicals. The formulation and laundering process can also affect the removal process. Encapsulated and wettable powder formulations were easier to remove than the emulsifiable concentrate.

Depending on the nature of the chemicals and the fabric type, one washing cycle may not be sufficient to remove the chemicals. Hence, multiple washing cycles are needed to effectively remove the chemicals. The time frame between the contamination and washing also affects the efficiency of cleaning. Immediate washing of the contaminated clothing can significantly improve the chemical removal. The storage of the clothing for a longer period can help the chemical to be strongly adhered to the substrate, hence making the removal difficult.

It is essential to remove the contaminants present on the surface or inside the matrix, or on both before the reuse. The chemical PPC can be cleaned by the combined process of pre-soaking, air drying, washing and drying at an elevated temperature. The use of appropriate chemicals is also essential for this purpose. The efficiency of decontamination can be calculated using the following formula:

$$\text{Decontamination } (\%) = (\text{weight loss} / \text{weight gain}) \times 100$$

where, weight gain = weight of the exposed specimen − weight of the virgin specimen

Weight loss = weight of the exposed specimen before decontamination

− weight of the exposed specimen after decontamination

The following factors affect the decontamination efficiency:

- The type and concentration of detergent: depending on the nature of the toxic chemical, the efficiency of a detergent will change. For example, pesticide removal was more efficient with an anionic phosphate detergent [173]. The cationic surfactants are often used to increase the attraction of the pesticide to the target foliar material. With the increase in the concentration of the removal of the pesticide, it is more effective up to a certain point [174].
- Prewashed product: for prewashed products, the decontamination efficiency will be higher as the product provides surfactants with a higher dosage. For example, the decontamination efficiency was higher for methyl parathion, deltametrin, trifluralin and triallate [175,176].
- Quality of water: the presence of minerals and salts in the water may reduce the cleaning efficiency.
- Auxiliaries used in laundering: the auxiliaries (fabric softeners, starch, bleaches and laundry boosters) used in laundering can alter the chemical energy available in the refurbishment process. For

example the presence of starch can alter the wetting behaviour of the fabrics and hence the decontamination efficiency.

- Heat and mechanical agitation: the increased temperature in the wash cycle can help in increasing the decontamination efficiency. However, higher heat in the presoak or rinse cycle may not be very effective. Increased agitation, longer extraction time and increased extraction volume can increase the removal of toxic chemicals from PPC.
- Water volume: increased volume of water can help in increasing the decontamination efficiency.
- Material variable: factors such as fibre content, yarn and fabric structure, fabric finishes in addition to the process parameters as described above affect the decontamination efficiency. The irregularities in the fabric surface act as a sink for pesticides. It is difficult to remove the chemicals from these places. Finishes that repel water such as a soil-repellent finish reduce the chemical absorption and facilitate removal.

The protective clothing for chemical protection is selected not only on the basis of the level of protection but also on the basis of several important factors such as ease of care and maintenance. Table 2.7 indicates a guideline while selecting the chemical PPC including the care, maintenance, use and disposal at the end of life cycle.

Some chemical PPC requires special storage conditions such as being away from sunlight, ozone or moisture. The manufacturer's instructions should be checked for the proper care and maintenance in addition to the storage. Any violations from these conditions may void the warranty and alter the performance of the PPC. The detailed information on chemical PPC can be obtained from ASTM F2061-00 (standard practice for chemical protective clothing care and maintenance instructions).

The laundering process used for the decontamination of the PPC should be effective in removing the contaminants so that they are safe when worn. Although in many instances where the cloth looks clean, the residual contaminants in the cloth can be carcinogenic to the skin. The laundering process should be carefully selected depending on the nature of the chemical used. The method suitable for the removal of one chemical may not be appropriate for others. Highly concentrated chemicals may not be removed completely compared to diluted ones.

It is very difficult to remove contaminants if the PPC is soiled with different chemical types or several PPCs with different contaminants are used in the same wash load. The PPCs treated with water repellent or a soil-release finish will be easier to clean when contaminated.

Table 2.7 Selection criteria for chemical PPC

Checklist	To Consider (example)	Checklist	To Consider (example)
Assessment of hazards and risks	Check MSDS	Skin care	After cleaning add cream to skin
Assessment of need of protection by developing a product specification	Tactility maintenance	Decontamination and cleaning	Air dry after cleaning in warm water
Determination of barrier material based on resistance data and usability	Neoprene	Storage	Avoid hot and cold temperature
Selection of the most appropriate Central Product Classification (CPC) product based on steps 1–3	Size 10, long sleeve	Inspection	Check for damage
Training of users	Knowledge about risks	CPC contaminated with hazardous materials	Dispose in accordance to regulations
Instruction for use	Warnings	Disposal of CPC in designated container	

2.6.4   Cleaning of other protective clothing

A surgical gown becomes contaminated with microorganisms during wear, which needs effective cleaning before being used again. After daily use or whenever the gowns become visibly soiled or wet by blood, body fluids or sweat, reusable surgical attire should be laundered in a facility-approved and monitored laundry. However, the laundering of surgical attire in home laundries is not recommended as it can result in the potential spread of contamination in the home environment. Surgical gowns simply worn in a medical environment are more likely to be soiled by per-spiration, body oils or material handled during the performing of other duties [177]. Soiled attires are not classified as contaminated unless the garments come in contact with blood or other potentially infectious mate-rials. Contaminated scrubs under no circumstances should be laundered in a domestic environment. However, all kinds of soiled attires can be laundered in a domestic environment.

In several countries, the rule states that "If the surgical gown (owned by the hospital or not) worn by the employee gets contaminated, the employer is responsible for laundering it" [178]. In addition, some of the hospitals specify the policy and procedures for launder-at-home situa-tions. There are four factors affecting the degree of decontamination of surgical gowns such as [179]:

1. The action of the washing chemicals and other aids that are used,
2. Washing temperature,
3. The dilution (repeated suds and rinse bath), and
4. The duration of the wash cycle.

The majority of the scrubs are P/C blends with the labels specifying "no chlorine bleach," which may affect the colour of the scrub. It is not uncommon to find the same restrictions on white items such as sheets, pillow cases and towels. Therefore, people concerned about a deter-gent's disinfectant capability include a new generation of bleach in its composition.

Many detergents commercially available for domestic use may not be appropriate for laundry sanitisers for hospitals. The laundry sanitisers for hospital outfits must demonstrate their efficacy against a representative gram-positive bacterium (e.g., *Staphylococcus aureus*) and gram-negative bacterium (e.g., *Klebsiella pneumonia*). If desired, additional organisms may be tested and claimed. The sanitiser should be able to "kill 99.9% of bac-teria" [180].

Doty and Easter [181] investigated the effects of the care and main-tenance of various protective clothing by washing and drying. The technical textiles were antimicrobial, stain repellent, stain release, mois-ture management and ultraviolet (UV) protection materials. Garments were subjected to repeated laundering and drying cycles in a single load. The performance of garments was evaluated prior to and after laundering (a maximum of 20 cycles).

The results showed that that laundering and drying of various PPCs in one load did not have a significant impact on the performance. Although they were subjected to 20 wash and dry cycles, their chemical structure was not affected. It was assumed that the mixing of various functional textiles during washing and drying can affect the performance of each other within the first few cycles. However, this assumption was not right as no change in the performance and chemical structure was observed. In addition, no garment acquired functional characteristics of other garments in the load. However, there was a slight change to the appearance and dimensional stability.

The FR high visibility garments must be laundered separately in water lower than 60°C or dry cleaned either with perc or petroleum solvent [182]. The use of natural soap, hard water, bleaches, a long washing cycle, over drying, high wash temperature, starch, fabric softener and other additives should be avoided. The use of bleach can damage the clothing whereas starch and softener may reduce the performance due to their presence in the fabric surface. High water levels, soft water, short extract time, detergent with high surfactant and low alkalinity, thorough cold water rinsing and permanent press/low setting ironing is suitable for these protective clothes.

Heavily soiled garments with abrasive soils can be washed at 40°C at the beginning of the cycle to reduce the abrasion. The washing load and chemicals should be established to avoid fabric abrasion and the redeposition of soil. Short extract time can help to avoid wrinkles whereas tunnel finishing or ironing after the short washing cycle can improve the appearance. The clothing can be repaired for minor faults not affecting the integrity of the garment using similar materials either by heat sealing or sewing on patches.

The FR high-visibility rainwear should be hand washed or machine washed using cold water and a gentle cycle to retain the FR properties and the high-visibility. Abrasive cleaners or solvents should be avoided. These items should be kept away from bleaches, softeners and dry cleaning. They should be hang dried and not ironed.

The protective clothing for molten metal can be laundered (either by home or industrial equipment at low washing and drying temperatures) or dry cleaned. The use of hard water should be avoided as the metal salts in hard water can form insoluble deposits on the fabric surface, which can affect the protection level. Excessive deposits may serve as a fuel for fire if the garments are exposed to an ignition source. Tunnel finishing is not suitable for these garments as it may result in excessive shrinkage.

As a wide range of equipment and chemicals is used for cleaning, these items should be tested by in-house laundering for any adverse effects. Garments soiled heavily or with splash metal should be dry cleaned for higher efficiency. The use of natural soap, long washing cycle, high-wash temperature, starch and bleaches should be avoided. The use of bleach can damage the clothing whereas starch may reduce the performance due to its presence in the fabric surface. High water levels, detergent with high surfactant and low alkalinity, and thorough cold-water rinsing is suitable for these protective clothes. Hydro extraction should be done at low speed only for a short duration.

While laundering at home, use home laundry detergent for splash metal. Avoid tallow soap, starch and bleaches. Select permanent press or the gentle cycle with cold or warm water (maximum temperature 60°C). Use the low/delicate cycle in tumble drying and remove promptly.

Conditioning, tunnel finishing (if used) and pressing temperature should not exceed 120°C. Perc or petroleum solvent should be used for dry cleaning. The residual dry cleaning solvent or washing chemicals should be removed. The clothing can be repaired for minor faults not affecting the integrity of the garment using similar materials either by heat sealing or sewing on patches.

The protective clothing used for sun/UV protection is prepared from the blends of cellulosic fibres and synthetics. The laundering of new clothes can improve their protection level, especially the clothes made of natural fibres. This might be due to the shrinkage that reduces the gaps in the structure. As the clothes become older they may offer decreased protection due to regular care. The effect the first wash has on the fabric's sun protection factor (SPF) is very crucial as it can account for most of the loss of SPF after the wash [183]. Hence, these clothes can be treated with UV absorbers to absorb more UV radiation.

While using sunscreen or lotions on the body, they should be applied 15–25 minutes prior to putting the UV clothing on to avoid staining or discoloration of the fabric. The UV-protective clothing should be washed soon after each wearing. If it cannot be washed, at least a rinse in cool water is necessary. While using the washer, cold water and a gentle washing cycle using mild soap/detergent can provide good results. These clothes should not be wrung, but rather laid flat to dry. The garments can be rolled in a dry, clean towel, to absorb excess water, which can speed up the drying process. They should not be bleached, dry cleaned or ironed, and should be stored when they are completely dry.

The cold-weather protective clothing needs special care during the care and maintenance as it contains multiple layers of different materials [184]. The appropriate care procedure for one layer may be harmful and/or ineffective for the other layers. The situation is aggravated if the cold-weather clothing is highly contaminated, which is often the case in some sectors. Heavily soiled or contaminated cold-weather clothing cannot be properly cleaned in cold water, especially oil-based dirt. In oil-based soiling, the oils are not sufficiently softened to be removed by the roll-up mechanism. Enzymes that are effective in cold water are an important additive to break down other insoluble dirt residues.

While cleaning the cold-weather protective clothing, the garments should be checked for any mechanical damage and repaired. The dirt and stains should be pretreated before washing or dry cleaning to facilitate their removal. Cold-weather protective clothing can be wet cleaned at a commercial laundry instead of dry cleaned with organic solvents. The professional wet cleaners use wet cleaning equipment with water and they operate in tandem with their dry cleaning machines. Wet cleaning the cold-weather clothing in a commercial laundry is more appropriate than using the organic solvents. The wet cleaning can be done on very dirty clothing, rain wear and items with microporous structures.

The outer layer of a cold-weather garment is soiled by air pollutants (carbon black, acidic gases), body excretions and direct contact with dirt or food residues. Among these, about 40% soil is water-soluble and 10% is solvent-soluble. Various textiles were grouped by Wentz [185] according to their preferred method of cleaning into two categories: aqueous and non-aqueous cleaning. Cold-weather items such as overcoats, parkas, rain-coats, sweaters, windbreakers, blankets and sleeping bags, were nearest to the aqueous end of the scale. After wet cleaning, the garments should be thoroughly rinsed for detergent residues, which are not desired.

In some occupations, the contamination of cold-weather work clothing is inevitable (e.g.,workers in many parts of the oil and gas sectors working in cold climates). Contaminants such as dirt and greases accumulate in the cold PPC, which reduces the effectiveness of its protection. Hence, the outer garments such as parkas need to be cleaned before the contaminants settle into the fabric, which makes them difficult to remove. The members of the U.S. military are advised about the difficulty of removing grease and oily contaminants from cold-weather parkas. These contaminants are hard to remove as high heat is needed to remove these stains, which may damage the parka's tape. In addition, many outer garments prepared from synthetics are also hard to clean at high temperature.

Aeration is a suitable approach for the removal of some types of con-taminants. The dry cleaning and professional wet cleaning is not always accessible to the workers working in the remote areas. The research of Crown et al. [162] is aimed at establishing care procedures for workers at remote locations, which are close to domestic laundry conditions. The use of laundry pre-treatments (with domestic pre-laundry sprays or in a degreaser) was necessary to remove motor oil from aramid fabrics. As the contaminants or number of laundry cycles increase, the level of difficulty increases to remove the contaminants.

The outer layer of the cold weather PPC with a durable water repellent (DWR) finish (silicone or fluoropolymer) may be removed due to multiple cleaning or may become ineffective due to dirt, detergent residues or fabric softeners. Hence, in order to get back the required protection it is necessary to reapply the DWR finish, which is available in sports outlets or outdoor clothing retailers.

The cold weather apparel prepared using Gore-Tex breathable water repellent membrane should be cleaned in accordance with its instructions for safe cleaning. For grease or oil-based stains a pre-wash spray before cleaning can provide improved results. Warm machine washing (40°C) using liquid or powder detergent without bleach or softener is ideal for cleaning. To remove the traces of detergent residue, it should be thoroughly rinsed. Although commercial dry cleaning can be used for a Gore-Tex membrane, special care should be taken by the dry cleaner to avoid any damage. The microporous structure of the breathable membrane may be clogged with the dry cleaning solvent as it contains surfactants and other additives. It should be treated with an extra rinse cycle with fresh solvent.

Furthermore, the polyurethane (PU) coating can be adversely affected by the cleaning solvent used in dry cleaning. In addition, PU coating is also damaged by the light due to the action of the UV. Hence, storage should be done away from the light. When there is a need to iron, a steam iron can be used by using a press cloth between the garment and the iron. The heat applied during pressing or even tumble drying helps in redistributing the finish in the fabric.

The insulating layer in the cold protective clothing is prepared from polyester or polyolefin fibres, which are sensitive to heat. If they are exposed to temperatures higher than their thermal transition temperature or melting temperature during tumbling or pressing, the excessive heat can cause shrinkage or damage to the material. Combined heat and pressure during pressing can reduce the insulation properties of the protective clothing.

Similarly, polyolefin non-woven scrims, baffles or tapes are used to secure insulation materials. Excessive heat can cause puckering and distortions to the garment due to shrinkage or melting. These components can face problems in tumble drying, dry cleaning or pressing. The non-woven insulating layer can also be physically damaged (e.g., torn or shift) by the mechanical action during the cleaning process. Hence, a gentle cycle can prevent this.

In some cold-protection clothing, wool fibre batts are used as the insulating layer. If the wool fibres are not treated for a shrink-resistant finish, they may shrink or stiffen in washing. This type of garment containing wool fibres should be dry cleaned or wet cleaned with reduced agitation.

The underwear used with cold-weather clothing is generally single layer. Hence, it is much easier to clean compared to the multilayered cold-weather apparel. As the underwear is in contact with the skin, it is essential to properly clean it. Generally, the underwear used with cold-weather apparel consists of long-sleeved tops and long-johns prepared from natural fibres such as cotton, silk and wool or synthetic fibres such as polypropylene or polyester or blends of both natural and synthetic fibres. It is easier to clean synthetic fibre-made undergarments than the natural counterpart.

The underwear should be cleaned frequently or after each wear to avoid the problem of body odour created by the absorption of sebum and apocrine sweat. The underwear containing synthetic fibres emits stronger odour generated by body fluids compared to natural fibres [186].

Socks used for cold-weather protection are layered to increase thermal insulation instead of single thick socks [187]. Repeated laundering after use can change the thermal insulation of socks. The initial washing of a new pair of socks can increase the thermal insulation due to increased thickness. However, repeated laundering can reduce the insulation due to significant fibre loss and decrease in the thickness. As the presence of moisture can reduce the thermal insulation, socks should be completely dried after laundering [188]. Manufacturers should specify commercial wet cleaning on care labels, which should be strictly followed. Some manufacturers recommend professional dry cleaning as laundering at home may be unsafe due to the water-sensitive components, dyes and finishes. The laundry variables (such as water/garment/detergent ratio, water pH and hardness, water temperature and rinsing cycle) should be carefully controlled in order to remove the contaminants effectively without redeposition.

2.7   Effects of cleaning on clothing properties

Both the wet and dry cleaning processes act as a degradative agent for garments that are routinely cleaned. A number of properties such as aesthetics, physical and mechanical properties are changed due to cleaning. The fibres and yarns in the fabric can be damaged by laundering, which can be difficult to attribute to a specific cause, as a number of mechanisms (e.g., mechanical agitation, water, detergent, temperature, cycle duration and drying method involved) are involved in this. Various effects of cleaning clothes are discussed in the following sections.

2.7.1   Effects of wet cleaning

Garments may sometimes fail during use because of the loss of strength of the yarns and fabrics due to use and maintenance. A garment is subjected to various tensions during wear and various chemicals, heat and agitation during washing and drying. The combined effect of tension, chemicals, heat and agitation can alter the properties of the clothing, which can be realised by the change in the shape (shrinkage or stretch) and colour [189]. Various damages caused to clothing are discussed in the following section.

The exposure of some clothing to direct or indirect sunlight may cause deterioration in the fibres [190]. The rate of deterioration will vary depending on the fibre content, yarn and fabric construction and the type of dyeing, printing and finishing applied to the fabric. Another significant cause is weathering, which is the cumulative effect of daylight, temperature, humidity, rain, abrasive dust, reactive gases (pollution) and cosmetic radiation on fabric. Moisture in the air and grime present in atmospheric acid fumes may reduce the tensile and tear strength of fabrics. Bleaching agents such as hydrogen peroxide (H_2O_2) or sodium hypochlorite (NaOCl) may convert cellulose (in cellulosic fabrics) into oxycellulose, which is much weaker than cellulose. If the bleach is not thoroughly rinsed out, it may cause damage to other fabrics with which it comes in contact.

Laundering effects include the loss of tensile strength, discolouration, overall change in appearance, breakdown of molecular structure and a change in the oxidation state or degree of polymerisation [15,17,25,137,191–195]. Hurren et al. [196] reported that the mechanical and chemical degradation of fabric during laundering is mainly due to the abrasion of wet fabric and cleaning agents, respectively. It is reported that DP-finished fabrics retained a higher proportion of their initial strength in repeated laundering when compared to untreated fabrics [197]. Lau et al. [198] studied the effect of repeated laundering on the performance of wrinkle-free-treated garments. They reported that wrinkle-free treatment can reduce the adverse effects of washing on mechanical properties.

The mechanism of breakdown is substantially the same for all fibre types, i.e., cotton, wool, silk, linen and rayon. The cause of the breakdown of a large proportion of the fibres is the transverse cracking that occurs at the position of maximum weakness in the structure as a result of flexing and bending stress suffered during wear. The surface fibres that are held lightly undergo gentle abrasion. The breakdown of cotton fibres under abrasive forces in different conditions (dry and wet) during normal wear and laundering has been investigated [199]. In the dry state, the surface layers are rubbed and eroded with no indication of fibrillar structure. In the wet state, the fibres swell, the fibrillar structure is loosened and the fibrils can be torn out from the fibre surface.

Murdison and Roberts [200] studied the damage done to cotton fabrics in laundering and storage by measuring the change in tensile strength and fluidity. Samples laundered in 1940 and re-tested in 1948 showed a lower tensile strength and a higher fluidity due to ageing. Uneven cracks also developed in the fibres. Fi Jan et al. [201] investigated the influence of laundering on the properties of cotton fabrics. It was found that the higher concentrations of hydrogen peroxide used at higher temperatures in a longer laundering cycle with lower liquor ratios result in a higher degree of chemical and mechanical damage.

The failure of seams in a sewn garment occurs due to the unsuitable selection of sewing thread, stitch type or stitch density, too shallow a seam allowance or too tight a fit. Although the fabric in a garment may remain in good condition, a failure of the seams reduces serviceability. The failure may be due to slippage or to the seam strength [157,202–204]. In many instances, seam pucker can develop after the care procedures in the cloths due to differential contraction along the line of a seam caused by the tension from the thread of the seam or the yarns of the fabric. Seam puckering is a disruption in the original surface area of a sewn fabric that gives a woollen and wrinkled effect along the line of the seam in an otherwise smooth fabric [205,206].

Slippage is the condition in which a seam sewn in the fabric opens under load and may close on removal of the load, although it may also cause permanent deformation [205,207]. Seam slippage is a particular problem in fabrics with slippery yarns or an open or loose structure. It is associated with seam allowance, seam type and stitch rate. Tension in the fabric or rubbing of the garment may result in yarn shifting, which causes slippage. Seam strength is the force required to break the sewing thread at the line of stitching.

Snagging: Snagging is the pulling out of warp or weft threads in a woven fabric, and wale or course threads in a knitted fabric through the contact with rough objects, which leads to the formation of loops on the fabric surface [202]. Only the appearance of a garment is changed by snagging and its other properties are not affected. Snagging is observed particularly in filament-type fabrics, and in extreme cases, a single blemish may render an article unserviceable even though unsightly ladders do not necessarily ensue. Soft twisted yarn and loose fabric structure are prone to snagging, which may rupture the yarn and ruin the fabric. Woven fabrics with long floats and fabrics made from bulked continuous filament yarns are susceptible to snagging.

Pilling: Pilling is the appearance of small bunches or balls of tangled fibres on the surface of a fabric, which are held in place by one or more fibres and give the garment an unsightly appearance [208]. Before the invention of synthetic fibres, pilling was mainly observed in knitted woollen items made from soft twisted yarns. Both woven and knitted fabrics are prone to pilling. The propensity may be related to the type of fibre used in the fabric, the type and structure of the yarn and the fabric construction [209]. Generally, pills are formed in areas that are especially abraded or rubbed during wear and can be accentuated by laundering and dry cleaning. The rubbing action causes loose fibres to develop into small spherical bundles anchored to the fabric by a few unbroken fibres [210].

Fabric made from natural fibres is less prone to pilling as the fibres break away and shed the pills. In synthetic fabrics, because of a higher strength of the fibres, they remain attached to the garment and accumulate to form pills. Pilling is particularly associated with nylon or polyester as may be seen in the collar of men's woven shirts made from P/C or nylon/cotton blends [211]. Woollen-knitted garments with a loose fabric structure made from soft twisted yarn (e.g., jumpers and cardigans) also suffer frequently from pilling [212]. This can be reduced by diminishing the migratory tendency of fibres from constituent yarns in the fabric and is achieved by the use of a higher twist in the yarn, reduced yarn hairiness, longer fibres and increased inter-fibre friction, a greater number of threads per unit length, brushing and cropping of the fabric surface and specialised chemical finishes [213].

The effects of fabric softeners and cellulase-enzyme containing laundry detergents on pilling were investigated [214]. It was observed that some softeners were not associated with an increase in pilling and that cellulase-enzyme detergent additives significantly reduced the amount of pilling on all cotton fabrics, except cotton interlock knits.

Abrasion: Abrasion is a progressive loss of fabric caused by rubbing against another surface. It has also been reported to occur through molecular adhesion between surfaces, which may remove material. The hard abradant may also plough into the softer fibre surface. The breakage of fibres has been reported to be the most important mechanism causing abrasion damage in fabrics [215]. Abrasion can be of three types: flat or plane, edge and flex. In flat abrasion, a flat part of the material is abraded; edge abrasion occurs at collars and folds; and flex abrasion rubbing is accompanied by flexing and bending. Abrasion is a series of repeated applications of stress. The selection of suitable yarn and fabric structure can therefore provide high abrasion resistance [216].

Abrasion resistance is dependent on several factors such as the fibre type and properties, yarn structure, fabric construction and type and the type and amount of finishing material present. High elongation, elastic recovery and the action of fibre rupture are more important than high strength for good abrasion resistance. Nylon fibre is considered to possess the highest degree of abrasion resistance while viscose and acetates have the lowest [211]. Polypropylene and polyester fibres also have good abrasion resistance. The abrasion resistance of wool and cotton can be increased by blending with nylon or polyester. Longer and coarser fibres help to improve the abrasion resistance of a fabric. Increased linear density and balanced twist in a yarn give the best abrasion resistance.

Laundering may cause significant abrasion in fabrics, thus shortening the wear life of a garment. Cotton fabrics laundered in hard water suffered significantly more edge abrasion than those laundered in soft water, and carbonate detergents caused more abrasive damage than phosphate detergents [217]. Neither detergent harmed fabrics when used with soft water.

A fabric with evenly distributed crimp between the warp and weft gives good abrasion resistance as damage is spread evenly between the threads. The higher abrasion resistance of fabrics with higher float (such as twill, satin and sateen) may be attributed to the easy relative mobility of threads, which helps in absorbing stress. This also is the cause of higher abrasion resistance in knitted fabrics, which have looser structures. Fabrics with optimum sett produce the best abrasion resistance.

If the fabric structure is too tight, it prevents the movement of threads, which are then unable to absorb the distortion. This results in lower abrasion resistance. A tight structure also causes the fibres to be stressed and fatigued beyond their yield point, which leads to breakage. Abrasion damage and its related effects are likely to be a significant factor in determining the wear life of a garment during normal use [218–221].

Colour fading: One of the major problems garments face is poor colour fastness. A coloured item may encounter a number of agencies during its lifetime that can cause the colour either to fade or to bleed into an adjacent uncoloured or light-coloured item [222]. New garments may experience colour loss due to the removal of excess colour that was not adequately rinsed after dyeing. Colour loss can occur by the migration of weakly bonded dye molecules out of the fibre. Colour loss during washing will stain other materials and this will be influenced by the ratio of coloured to uncoloured items, fibre content of other items and end-use conditions [223,224]. A specific hue may be produced by the mixing of two or more dyes. If one component is degraded or lost from the material, the colour will be altered.

The type of dye, the particular shade used, the depth of the shade and the dyeing process all affect the fastness of a colour. Some coloured or printed garments change colour significantly during use. This may be caused by abrasion, rubbing, atmospheric conditions such as UV light, oxides of nitrogen or ozone, acid or alkaline substances, laundering or dry cleaning, ironing; perspiration, rain water, chlorinated water or sea water. Colour loss due to abrasion may be caused by localised wear such as rubbing the elbows against a desk, excessive mechanical agitation during washing or an attempt to remove a stain by rubbing. The exposure of a garment to direct or indirect sunlight may cause colour change resulting in fading because the UV rays in sunlight cause damage to the dye structure.

Atmospheric gas fading or fume fading is the colour change of a fabric caused by acid gases in the atmosphere that are formed in combustion processes. Garments left hanging for a long period of time will be affected by fume fading. Nitrogen dioxide (NO_2) is primarily responsible for gas fading. Ozone fading (O-fading) occurs in dyes that are colour-fast to fume fading if a high amount of ozone is present in the atmosphere. Disperse and direct dyes are more vulnerable to O-fading, and blue and red dyes are affected to a greater degree than others. Ozone may cause bleaching in acetate, cotton and nylon fabrics.

Colour loss may sometimes be due to acid or alkali present in various products. Certain dyes used in dyeing wool fabrics change colour when exposed to acidic conditions. Alkaline colour change is observed in dark blue and black acetate fabrics. Colour loss through laundering and dry cleaning has been investigated by several researchers [225–227]. Colour fastness to perspiration is also an important factor for consideration by manufacturers [228,229]. Perspiration is harmful as the bacterial action may lead to a loss or change of colour and finish, loss of fabric strength, odour problems, salt rings and deposits. Alkaline-sensitive dyes may be damaged by fresh (acidic) and decomposed (alkaline) perspiration. Many of the direct, basic, acetate and metallic dyes are affected by perspiration. Perspiration may change the hue, cause bleeding of the dye and staining of lighter areas.

Some fabrics change colour in rain water and chlorinated water, and some garments may change colour when worn near the ocean. This is normally observed in wool fabrics and is a result of the action of sodium chloride on the dye.

Shrinkage: A garment with dimensions that remained constant throughout its useful life would have great technical value. Shrinkage is a serious problem in different garments, originating from dimensional changes in the fabric when it is subjected to washing and dry cleaning. In recent years, this problem has become more prevalent due to the wide acceptance of casual wear such as tights, pants, blouses and sportswear. Several researchers have focused on various causes of shrinkage in woven [230–235] and knitted structures [15,137,236–238].

Most fabric production processes involve the application of high tension, which leaves strains in the fabric. These residual strains must be removed by the manufacturer before the fabric is converted into a garment or it will lead to shrinkage (i.e., relaxation shrinkage) when the fabric is washed [15,239,240]. Several washings are usually required for the complete relaxation of the fabric. The largest changes occur after the first laundering cycle, whereas they become stable only after six laundering cycles [241]. The residual strains are relaxed by the hot and wet conditions of washing, which can lead to shrinkage. Other forms of shrinkage include hygroscopic or swelling shrinkage, felting shrinkage, thermal or heat shrinkage and progressive shrinkage.

Hygroscopic or swelling shrinkage is caused by the swelling of fibres due to the absorption of moisture [242]. This shrinkage is highest in the rib weave of wool, cotton, rayon, acetate fibres or their blends and occurs when the garments are washed or dry cleaned. This group of fabrics may be pre-shrunk and should be cleaned in solvents with low relative humidity (RH). Felting shrinkage is related to wool and other hair fibres and the effect is closely related with their scaly surface features. It is accentuated if woollen items (e.g., socks and underwear) require periodical washing [238,240]. Felting shrinkage may be caused by excessive mechanical action during washing, high-temperature drying and high RH of the solvent during dry cleaning.

Thermal or heat shrinkage occurs in clothing made of synthetic fibres when subjected to high temperature. The magnitude of the shrinkage depends on fibre morphology and the temperature, tension and washing time. Some knitted garments are made from heat-sensitive fibres that shrink excessively when exposed to heat during drying and finishing. This can be reduced by permanent heat-setting of the fabric during finishing. Progressive shrinkage is the relaxation and swelling shrinkage caused by successive cleaning processes.

Although shrinkage is a common phenomenon in both the length and width of a garment, some may shrink in one dimension and stretch in another. The effects of many aspects of home laundering on the dimensional stability and distortion of cotton knits have been analysed by several researchers [233,243–245]. Studies on cotton knits showed that tumble drying causes greater levels of shrinkage than line drying in the first few laundering cycles [246–248]. It was also shown that the level of shrinkage should increase with successive laundering cycles, reaching a maximum after 5–10 cycles.

Stretch: There are several accounts of knitted garments increasing in size during wet cleaning, dry cleaning and finishing. In some cases, a fabric that shrinks in length will stretch in width. The use of a garment may cause some fabrics to stretch out of shape. Some knitted garments may stretch out of shape if they are hung to dry while still dripping with water or solvent and some may stretch due to manipulation during steam finishing, as the fabric is warm and moist from steam. Stretching can be controlled by suitable yarn, fabric structure and appropriate care during cleaning.

Bagging is a stretch phenomenon commonly observed in knitted garments and occurs on the cuffs, ankles and collars [249]. This phenomenon is related to insufficient elastic recovery of the garment. Where a garment is made from fabric with poor elastic recovery properties, it will soon show a baggy appearance (e.g., at the knees of trousers and the elbows of jackets). Elastic fabrics containing elastane are capable of stretching far more than conventional fabrics, so increasing the elastic recovery. However, the use of higher amounts of elastane increases fabric stiffness and lowers the tensile and tearing strength [250].

Change of sensory and comfort properties: Fabric comfort properties are affected by the sensorial properties (i.e., the fell of the fabric) and the transfer properties (such as moisture vapour, heat and air) [158,251]. The agitation cycle in the presence of a chemical during washing can lead to the loss of sensorial as well as the transfer properties. For example, the loss of wool fabric sensorial properties was investigated by Mackay [19]. It was observed that the agitation of woollen items was the main reason for felting shrinkage, which led to the loss of sensory properties.

Similarly, the thermophysiological comfort properties of a garment, which is an essential property [252,253], can be affected by the laundering process. The effect of conventional and ultrasonic washing methods on the thermophysiological comfort properties of knitted fabrics (made from polylactic acid (PLA), cotton, polyethylene terephthalate (PET), and poly-acrylic (PAC)) was investigated by Uzun [254]. The fabrics were washed (by using conventional and ultrasonic washing methods) 10 times for 15 and 60 minutes under 40°C. The test results showed that the washing processes affected the thermal conductivity, thermal resistance, thermal absorptivity, water vapour permeability and heat loss. Fabrics washed for 15 minutes by ultrasonic showed significantly lower thermal resistance as compared to conventionally washed fabrics. The PLA fabrics exhibited high dry and wet thermal-resistance values before the washing processes, and washed PLA fabric had relatively lower thermal conductivity. Ultrasonically washed fabrics had higher thermal conductivity than conventionally washed fabrics. The washing times did not significantly affect the thermal conductivity of fabrics.

In addition to the problems listed above, wrinkling, tearing, microbial attack and environmental decay are also associated with garment damage. Garments most commonly become wrinkled during use and are unable to recover from folding deformation. Wrinkling also arises from laundering, especially in cotton fabrics, and causes particular concern to consumers. Wrinkle resistance in a fabric enables it to resist the formation of wrinkles when subjected to folding deformation. The wrinkle behaviour of a fabric depends on the type of fibre and yarn, the fabric construction and the finish applied to the fabric. The resiliency of wool and polyester fibres gives good wrinkle recovery. High twist yarns can improve the wrinkle resistance of a fabric. A tightly woven fabric (having more ends and picks) is more prone to wrinkling than one that is loosely woven. Thinner fabrics are also more prone to wrinkling. A plain woven fabric will wrinkle more than twill or a 4×4 basket weave. Different finishes can be applied to a fabric or garment to improve the resiliency and therefore the wrinkle recovery.

Tearing is a phenomenon commonly faced by loose-fitting garments. The resistance of a garment to tearing is measured by the force required either to start or to propagate a tear in the fabric. Tearing strength depends on the fibre type, the yarn strength and the fabric construction. Under the action of tearing, the threads in a fabric group close by sliding instead of permitting the successive breakage of individual threads. The grouping of threads becomes easier if the yarns are smooth and able to slip over each other. For this reason, twill and matt weaves exhibit better resistance to tearing than do plain weaves. Fabrics with higher end and pick density prevent the threads from grouping, which reduces the tearing strength. Different finishes, such as anti-crease treatments, may reduce the tearing strength.

Garments may also be damaged by microbial attacks such as insects, mildew and rot. Many insects, including moths, beetles, crickets, roaches and termites, will eat the fibres or any food matter that is allowed to dry on a fabric. Natural fibres are more prone to insect damage, though synthetic fibres may also be damaged in this way if they are soiled.

The most common form of insect damage is caused by moth larvae in wool and hair fibres. Such damage may not be noticed prior to cleaning and flexing of the fabric. During cleaning, the yarns may be weakened and may break at the point of attack, resulting in a hole in the garment. The risk of insect damage is lowered if garments are cleaned properly and are free from stains before storing. Moth-proofing finishes can also be applied to garments.

Vaeck [255] has investigated the chemical and mechanical wear of cotton fabric during laundering. Cotton fabric was laundered up to 50 times and the decrease in tensile strength taken as a measure of wear. The wear resulted from chemical degradation (caused by oxidising agents or bleaches) and mechanical abrasion. He found the tensile strength loss to be much lower with cold bleach than with hot. It was also reported that in Western Europe, the United Kingdom (UK) and the United States (U.S.), sodium hypochlorite is the most commonly used bleach in commercial laundries, while in Central Europe peroxides are preferred. Hypochlorite bleaching may be done either in a cold rinse or in wash liquor. Most European countries prefer cold rinse, while the UK and the United States use the second method.

The loss of durability in a garment may sometimes be due to acids or alkalis. Hydrochloric and sulphuric acids are extensively used in industrial plants, dental, medical, photographic, automotive batteries and in all chemical laboratories. Accidental contact with these products may cause fabric damage, i.e., strength loss, disintegration of affected areas and the appearance of holes. Alkaline damage (caused by caustic soda, caustic potash and strong alkaline washing compounds) affects silk, synthetic protein fibres, wool and other hair fibres. Caustic alkalis are common in many household cleaning aids. Sometimes acid or alkali damage will become evident after a garment is cleaned.

It was observed that the repeated laundering of cotton fabrics in alkaline solutions near boiling point and without bleaching left the fabric strength unchanged, although linen fabrics suffered 20–30% strength loss. After the Second World War, Parisot [256] investigated the mechanical wear of garments by measuring the bursting strength and chemical degradation that was expressed as a degree of polymerisation. Many researchers also reported chemical degradation in laundering, but did not address the loss of tensile strength caused.

Although DP finishes impart shape retention, dimensional stability and wrinkle recovery to cotton and P/C fabrics, other properties such as strength, extensibility and abrasion resistance are adversely affected [257–259]. The extensibility of a wrinkle-free-treated garment is significantly reduced after repeated laundering [198].

The laundering process destroys the forensic evidence in apparel, which was investigated by some researchers [260,261]. For example, the effect of laundering on the blunt force impact (BFI) was investigated by Daroux et al. [260]. This research investigated the BFI damage in common apparel fabrics and the effect of prior and post-laundering on this damage. Two single jersey fabrics (100% cotton, bull drill) were investigated as single and double layers using an impactor representative of a hammer face. The force transmitted through the specimens was measured and the impulse was calculated. To establish the effects of impact damage on new, dimensionally stable (laundered 6 times) and aged fabrics (laundered up to 30 times), and the effects of laundering on impacted specimens, impacting and laundering were completed cumulatively. It was observed that the BFI left recognisable damage patterns in the specimens, with a varying amount of damage. Both the visible and microscopic damage were altered after laundering. Prior laundering impacted fabrics produced holes in some specimens, and some fibres exhibited a failure characteristic of BFI.

Life cycle-assessment studies on garments have shown that the period for which a garment is in use is usually the most energy-demanding period during the product's life cycle. This is even higher than the energy needed in the production or transportation phases. Although the technology of laundering has undergone several changes, it is still influenced by the social, cultural and moral norms. The length of time a person wears a cloth before washing has not been reduced, but rather increased. The individual's wardrobe now contains more clothes that are cleaned more often than before [262]. Washing more often counteracts the technological improvements in laundry.

The consequences of changing the washing temperature, filling grade, detergent dosage or drying method on cleaning effect, energy and water consumption were evaluated by Laitala et al. [101] by laboratory-based tests. The results showed that the detergents used today are better suited for low-temperature washing. An efficient detergent of today can provide better results at 30°C than with a less efficient detergent of the past even at 40°C. While laundering only slightly soiled textiles or small loads, the detergent amount can be reduced. Many textiles washed at a higher temperature (60°C) suffered colour loss compared to lower temperature (40°C or lower). Line-dried textiles showed less shrinkage than tumble-dried textiles. The results can motivate consumers to do effective cleaning, which can reduce the environmental impacts caused by textile maintenance.

In several instances, a large portion of garments are washed purely habitually instead of examining the level of soiling [262]. Generally, clothing items such as underpants and T-shirts that are always in contact with the skin should be washed after each use [50,263]. The cleanliness of clothes can be assessed by consumers by evaluating the visible stains and odour before and after washing [264]. The acceptance level of cleanliness and body odours greatly varies over time. Body odours are considered appalling today, therefore, clean clothes, daily washes and use of artificial perfumes is almost a social norm [265]. Including the cultural influences, personal choices affect these hygiene practices.

In several instances, the complete washing of clothing can be replaced by other cleaning methods such as stain removal, airing or brushing. Airing can be effectively used for woollen materials or for garments that have a slight odour but are not soiled. An approximate estimation by Uitdenbogerd [266] revealed that using all clothing items one more day would save about 100 wash cycles per year in households (families with children). The energy used in washing clothes can be eliminated by the use of disposable clothing. However, this would increase the volume of production, distribution and disposable waste [267]. This option may be best suited for clothing used in specific areas, such as disposable gowns for the medical sector.

Depending on the nature of the fibre used in the clothing, the frequency of washing can be varied. For example, woollen garments can be washed less often than cotton garments due to inherent soil repellency of woollens, which can save energy. However, if the consumer is not aware of this fact or does not want to follow the recommendation, the potential saving is lost.

2.7.2   Effects of dry cleaning

The use of chemicals, heat and agitation during dry cleaning can also lead to the physical damage, dimensional instability, colour loss and many other problems as discussed in the laundering section. However, the degree of damage may be of a different extent depending on the clothing type and dry cleaning parameters. It is essential to check the garment is dry cleanable from the care label, or else there will be irreparable damage to the cloth.

Perc is the strongest solvent generally used for the cleaning of most of the textiles. The use of perc can lead to colour loss, especially at higher temperatures. Several times the combination of high temperature and a longer cycle can damage trims, special buttons and beads on some garments. Perc can also dissolve some adhesives and strip the plasticiser out of the polyvinyl chloride (PVC). Some dyes, finishes and resins are not colour fast enough to the effect of perc as they can lose their colour. Perc is better suited for oil- or grease-based stains (which only account for about 10% of all stains) than the common water-soluble stains (such as wine, tea, coffee, and blood). It is a well-known fact that the garment dry cleaned with perc leaves a mild chemical smell on garments.

Dry cleaning is not the answer to all soil and stain removal problems. Sometimes, stains become permanently embedded in the materials.

At times, the fibre or fabric cannot withstand normal cleaning and stain-removal procedures. Decorative trim may not be compatible with a dry cleaning solvent. Hence, it is important that consumers as well as dry cleaners read all care labels and follow the instructions.

2.7.3   Precautions during dry cleaning

A wide range of chemicals can be used for dry cleaning as discussed above. Heavily soiled/stained clothing should be pre-cleaned or spot-cleaned with the chemicals before being added to the dry cleaning machine. The types of chemicals should be carefully selected based on the type of the fabric and fibre. If the stains are not being removed properly after the dry cleaning, they can be spot cleaned again using the same chemicals used in the pre-cleaning. Generally there are three types of pre-cleaning or spot-cleaning chemicals commercially available, namely wet-side agents, dry-side agents and bleaches. The wet-side agents help to remove water-soluble stains from the fabric. These agents can be again subdivided into alkaline, acidic or neutral depending on their pH.

The alkaline agents such as ammonia, lye, potassium hydroxide (KOH), sodium hydroxide (NaOH) and protein-based home detergents are used to remove oil and protein stains. The acidic agents such as acetic acid (CH_3COOH), sulphuric acid (H_2SO_4), glycolic acid ($C_2H_4O_3$), hydrofluoric acid (HF) and oxalic acid ($C_2H_2O_4$) are used to remove tannin or plant-based stains. The neutral agents such as neutral synthetic detergents and surfactants can be used to remove water-soluble stains, water-soluble dyes, food and beverages.

The dry-side agents are used to remove oily stains, waxes, greases, fats, paints and cosmetics. These agents are based on non-aqueous solvents (acetone (CH_3COCH_3), carbon tetrachloride (CCl_4), perc (C_2Cl_4), trichloroethylene (C_2HCl_3), methylene chloride or dichloromethane (CH_2Cl_2), amyl acetate ($C_7H_{14}O_2$), etc.), alcohols (ethanol (CH_3CH_2OH), methanol (CH_3OH), isopropyl alcohol (C_3H_8O), etc.) and petroleum-based solvents. This group of chemicals involves some of the most toxic materials used in dry cleaning.

Bleaches are used if all the stain-removal techniques have failed. The bleach should be used with precise care to avoid the colour fading. They are also used in traditional laundering operations. The bleaches can be grouped into oxidising bleaches (hydrogen peroxide (H_2O_2), sodium perborate ($NaBO_3 \cdot nH_2O$), sodium hypochlorite (NaClO) and sodium percarbonate ($2Na_2CO_3.3H_2O_2$)) or reducing bleaches (oxalic acid ($C_2H_2O_4$), sodium hydrosulphite (NaHS), titanium sulphate (O_8S_2Ti) and sodium bisulphite ($NaHSO_3$)).

chapter three

Equipment for the care of textiles

Various equipment involved in the care of textiles includes washing, drying and pressing equipment. Washing equipment, or washers, is designed to wash, rinse and extract water from clothes and make provisions for setting the time, temperature and volume of water. The three major functions of washers are: (1) removal of soiling from the clothes, (2) the rinsing of soap or detergent and soil from the wash process and (3) the extraction of most of the wash and rinse water prior to drying. Drying equipment, or dryers, is used to dry the clothes, which can sometimes be combined with the washing equipment. Pressing equipment is used to press the garments to improve the aesthetic appeal of the clothes.

3.1 Washing equipment

A washing machine, or washer, is a machine to wash laundry, such as clothing and sheets. The term is mostly applied to equipment that uses water as the solvent compared to dry cleaning (using an alternative solvent) or ultrasonic cleaning. Washing equipment can be classified as top- or front-loading.

3.1.1 Top-loading washing machines

Top-loading machines are fitted with an agitator that transfers mechanical energy from the motor to the clothes [30]. The mechanical agitation, combined with detergent, removes soiling and keeps it in suspension. The central cylinder (solid or perforated) that contains the clothes to be washed is usually steel coated with porcelain enamel. As the detergent cycle is completed, the water is extracted from the fabric through the perforations of the cylinder. Front-loading machines have a cylinder but no agitator. The interior of the cylinder consists of baffles that lift the clothes from the water and then drop them back into it.

The water level will depend on the load size and the design of the equipment. Top-loading machines use more water for each cycle, but have fewer cycles than front-loading machines. The temperature of the water ranges from cold to 60°C and is controlled by a temperature unit fitted to the control panel of the washer. In some cases, cold-water washing can be used, thus conserving energy. Cold-water washing is used for knitted polyester or nylon and permanent press items that wrinkle in hot water. Following the washing cycle, the garments are spun dried or put through a wringer while still hot. This reduces shrinkage, especially in knitted fabrics, chino pants and some non-sanforised items. It also prevents the setting of certain types of stains (such as milk, egg and blood, which may become permanent if washed in hot water). Hot-water washing should be used for the removal of grease and oil stains.

Top-loading washing machines are so named because they allow adding the clothes in the tub vertically without having to bend down. They are more popular for domestic applications in New Zealand, Canada, the United States, Australia, Latin America and Asia. They have several advantages including the fact that clothes can be added even after starting a washing cycle. They can also use a wide variety of detergents and bleaches. The price of this equipment is cheaper than the front-loaders. The advanced top-loading washers can include special features in addition to the basic washing cycles. Figure 3.1 shows a top-loading washing machine with various components, which may slightly vary depending on the manufacturer. The detailed description of each component and function of the washing machine is not within the scope of this book. However, a brief description of the principle of operation is described below.

In top-loading machines, the clothes are placed in a vertically mounted perforated cylinder within a water-retaining outer tub. A finned water-pumping agitator is centrally located in the bottom of the basket. Clothes are loaded through the top of the machine, which is covered with a hinged door. While the washing cycle is started, the outer tub is filled with a sufficient amount of water to fully immerse and suspend the clothing freely in the cylinder. The agitator movement pushes water by the paddles towards the edge of the tub, which in turn returns the water towards the agitator to repeat the process. The rotational direction of the agitator is periodically reversed, as continuous rotation in one direction would just spin the water around the cylinder with the agitator rather than the water being circulated in a torus-shaped motion. Some washers supplement the water circulation with a large rotating screw on the shaft above the agitator, which helps the water move downwards in the centre of the cylinder.

Overloading a top-loading washer can either jam the motion of the agitator or damage the motor or gearbox, or even tear the fabrics. Extreme overloading can lead to the clothes being wrapped around the agitator shaft, restricting their motion and jamming the fabric. The use of pre-wet clothes in a washing load can restrict water circulation, resulting in poor cleaning.

Energy-saving washing units are preferable as consumers adapt to a greener lifestyle [268]. Many companies have incorporated such features

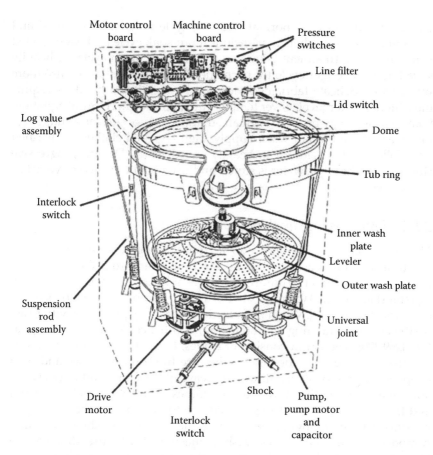

Figure 3.1 Top-loading washing machine.

on their products with eco-friendly wash cycles. For domestic laundering, the temperature of cold water can vary from extreme cold to water at body temperature. However, a water temperature of 27°C or more gives the best results. Special detergents that dissolve readily and have good cleaning properties have been developed for use in cold water. However, cold-water washing will not be bacteria free. Cold-water washing followed by tumble drying at 70°C will result in little or no bacterial contamination. Investigations by Witt and Warden [268a] showed that the important factors for preventing bacterial growth are water temperature (between 50°C and 60°C), detergent and longer washing cycles. Washing followed by steaming or ironing on a hot press has been shown to provide adequate disinfection.

The duration and degree of mechanical agitation in a washing cycle depends on the load, the amount of water used and the extraction of water from the load. A normal washing cycle may be between 30 and 38 minutes. Some washers may include a pre-soak cycle for heavily soiled items requiring treatment with enzymes, a super wash cycle for heavily soiled items, a longer cycle for permanent press items that require more water and a delicate fabric cycle for delicate items and blankets requiring a higher water level, lower temperature and low-mechanical agitation. Some washers may be fitted with an auxiliary device such as a suds saver for areas with a limited water supply. Lightly soiled clothes are washed first, followed by medium to heavily soiled items. During progressive rinse cycles, the rinse water is stored and used again in successive detergent cycles.

3.1.2   Front-loading washing machines

The construction and working of a front-loading washing machine is identical to a top-loading washing machine apart from the fact that it contains a cylindrical drum (tumbler washing tub) in place of the long agitator (Figure 3.2). Washing is performed by tumbling the clothes in the cylindrical drum [30]. When front-loaders spin, they permit gravity to act on the clothes (tumbling and bouncing) and hence do not require agitators [269,270]. The drum consists of blades, termed as agitating vanes or paddles, on its upper side. The clothes are lifted up by vanes and then dropped by gravity, which flexes the weave of the fabric and forces water and detergent solution through the clothes. The rotation of the drum and the blades generate strong currents in the water, and the rubbing of clothes assists in removing the dirt from them. Figure 3.2 shows the inner components of a front-loading washing machine and Figure 3.3 shows the basic principles of the working of a front-loading washing machine.

The front-loading designs are widely used in Europe [271], although they are also found in the Middle East, Africa and Asia. Most industrial washers around the world are front-loaders. As the washing process does not require the free suspension of the clothing in water, only enough water is needed to wet it. A reduced amount of water means the use of less soap, and the repeated dropping and folding action of the tumbling produces large amounts of foam or suds. A number of semi- or fully automatic washing machines are available comprising a heater in the bottom of the washing tub to generate warm water for loosening and stripping dirt particles from the clothes for rapid cleaning.

3.1.3   Top- versus front-loading washing machines

There are some differences between the top- and front-loading washing machines [272]. The mechanical and electrical design features are simple in front-loading washers. A top-loading washer keeps water inside the

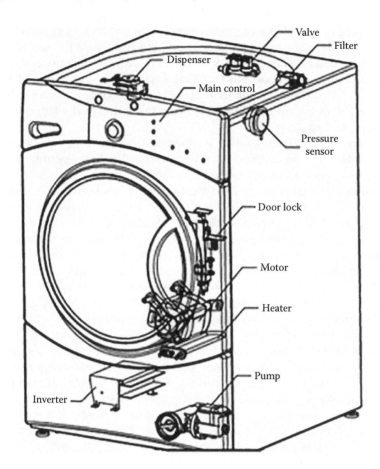

Figure 3.2 Front-loading washing machine: inner components.

Figure 3.3 Principle of the working of a front-loading machine.

cylinder, where the force of gravity acts down on the water; whereas, in a front-loading washer, water should be held tightly by the sealed door to prevent water leakage. This door is locked during the entire washing cycle, as opening the door could result in water gushing out onto the floor. In front-loaders, there is a chance that the clothes could be accidentally cut between the door and the drum, resulting in tearing of the fabric during tumbling and spinning.

Almost all the front-loading washers for the consumer market must use a folded flexible bellows assembly around the door opening to hold the clothing contained inside the basket during the wash cycle. Without the bellows assembly, small clothing items such as socks could slip out of the wash drum near the door and fall down the narrow slot between the outer tub and inner basket, plugging the drain and possibly jamming the rotation of the inner basket. Retrieving these items from the outer tub and basket can require complete disassembling of the front of the washer, including the inner wash basket. Commercial and industrial front-loaders usually do not use the bellows. They place all the small objects in a mesh bag to prevent these items from being lost near the door opening.

Clothing can be packed more tightly in a front-loader, up to the full drum volume (while using the cotton wash cycle), compared to top-loading washers. This is because wet clothes usually fit into smaller spaces than dry clothes. Front-loaders are able to self-regulate the water needed to achieve correct washing and rinsing cycles [268]. Overloading of front-loading washers pushes the clothes towards the small gap between the loading door and the front of the wash basket, potentially resulting in clothing lost between the basket and outer tub. In severe cases, the tearing of clothing and jamming of the motion of the basket can occur, resulting in the malfunctioning of the washer.

There are some differences in terms of cleaning efficiency and water usage, which are discussed below.

Cleaning efficiency and water usage: The use of energy, water and detergent is lower in front-loaders compared to the top-loaders [268]. However, the duration of a washing cycle is longer and is often computer-controlled with additional sensors to adapt the wash cycle to the needs of each load. As this technology improves, the human interface will also improve, which will make it easier to understand and control the different cleaning options. Front-loading washers usually use less water than top-loading residential clothes washers. Estimates show that front-loading washers use only one-third to one-half of the water used by top-loaders.

Spin-dry effectiveness: Front-loading washers can spin at much higher speeds (up to 2000 rpm), while top-loading washers (with agitators) can go up to 1200 rpm. High-efficiency top-loaders with a wash plate (instead of agitator) can spin up to 1400 rpm, as their centre of gravity is lower. Higher spin speed removes higher amounts of residual water, making clothes dry faster. This also reduces time and energy of drying, if clothes are dried in a clothes dryer. However, there is also much more risk of clothes being damaged at higher speeds.

Cycle length: The cycle time of top-loading washers is generally shorter, as their design focuses on simplicity, greater performance and speed of operation [273].

Water leakage: Top-loading machines are less prone to water leakage, as gravity cannot drag the water out the loading door on top as it can in the case of front-loading washers. Front-loading washers require a flexible seal or gasket on the front door, and the front door must be locked during the washing cycle to prevent it from opening and having a large amount of water spill out. This seal may leak and require replacement frequently. However, many current designs of front-loaders use very little water so they can be stopped mid-cycle for the addition or removal of clothes without water spilling out.

Wear and abrasion: Top-loading washers require an agitator or impeller to force enough water through the clothes to clean them effectively. While impellers may be very rough especially on bigger loads, agitators greatly decrease mechanical wear and tear on fabrics. On the other hand, front-loaders use paddles in the drum to pick up and drop clothes repeatedly into water for effective cleaning. Hence, during a washing cycle, clothes frequently rub against each other and this action causes more wear.

Noise: Front-loaders operate at a lower noise level compared to top-loaders as the door seal helps in preventing the noise. Top-loaders usually need a mechanical transmission, which can generate more noise than the rubber belt or direct drive found in most front-loaders.

Initial cost: In countries where top-loaders are preferred, buying front-loaders can be more expensive. However, their lower operating costs can ultimately lead to lower total cost, especially if the cost of energy, detergent or water is expensive. Similarly, in countries where front-loaders are the choice, top-loaders can be more expensive than basic off-brand front-loaders, although without many differences in the total cost of ownership apart from design-originated ones. In addition, manufacturers have tended to include more advanced features such as internal water heating, automatic dirt sensors and high-speed emptying on front-loaders, although some of these features could be and are implemented on top-loaders.

Compactness: A front-loading washing machine uses much lower space than top-loaders. These models can save space in homes with limited floor area, since the clothes dryer may be installed directly above the washer in a stacked configuration.

Maintenance and reliability: Top-loading washers may not need a regular cleaning of door seals and bellows. The potentially problematic door-sealing and door-locking mechanisms used by front-loaders are not needed in top-loaders. On the other hand, top-loaders use mechanical gearboxes that are more vulnerable to wear than simpler front-load motor drives.

Accessibility and ergonomics: Front-loaders are more convenient for very short people and those with paraplegia, as the controls are front-mounted and the horizontal drum eliminates the need for standing or climbing. For people who are not unusually short, top-loaders may be easier to load and unload, since reaching into the tub does not require stooping. Risers also referred to as pedestals, often with storage drawers underneath, can be used to raise the door of a front-loader closer to the user's level.

3.1.4   Other designs

In addition to the two washing machine designs, there are also several variations of them. Impellers are used instead of agitators in top-loading machines in Asian countries. Impellers are similar to agitators, but they do not have the centre post extending up in the middle of the washing cylinder.

Some top-loaders are similar to front-loading machines. They have a cylinder rotating around a horizontal axis as the front-loaders do. However, there is no front door and a top lid with a hatch (that can be latched shut) provides access to the cylinder. Clothes are loaded from the top and the hatch and lid are closed. The machine operates and spins similar to a front-loader. These types of machines usually have a lower capacity, are narrower but taller than front-loaders and are intended for use where only a narrow space is available. These machines are sometimes used in Europe. The advantages of this type of machine are: they can be loaded without bending down; they do not require a perishable rubber bellows seal and instead of the drum having a single bearing on one side, it has a pair of symmetrical bearings (one on each side), which avoids asymmetrical bearing loading and potentially increases the life of the machine.

There are also combo designs that combine the complete washing and drying cycle in the same machine, eliminating the transfer of wet clothes from a washer to a dryer. These machines are better suited for overnight cleaning as the combined cycle is longer. However, the effectiveness for cleaning larger batches of laundry is drastically reduced. The drying process uses higher energy than the two devices in isolation, as a combo machine not only must dry the clothing, but also needs to dry out the wash chamber itself. These machines are preferred in Europe, as they can be fitted into small spaces.

3.1.5 Dry cleaning machines

Although there are various makes/models of dry cleaning machines, they all work on a similar principle. Essentially a dry cleaning machine consists of four basic components:

1. Holding or base tank,
2. Pump,
3. Filter, and
4. Cylinder or wheel.

The holding tank is a reservoir for the dry cleaning solvent. A pump is used to circulate this solvent through the machine during the cleaning process. Filters are used to trap solid impurities. A cylinder or wheel is where the garments are placed in to be cleaned. The cylinder has ribs to help lift and drop the garments.

The operation of the dry cleaning machine is simple and straightforward. The solvent is drawn through the filters to trap any impurities from the tank by the pump. The filtered solvent then enters the cylinder to flush soil from the clothes. The solvent leaves the cylinder button trap and goes back to the holding tank. This process is repeated several times during the cleaning cycle, ensuring effective cleaning.

After the cleaning cycle, the solvent is drained and an extract cycle is run to remove the excess solvent from the clothes. This solvent is drained back to the base tank. During extraction, the rotation of the cylinder increases in order to use centrifugal force to remove the excess solvent from the clothes.

After extraction is complete, clothes are either transferred to a separate dryer or on most machines, dried in the same unit by a closed system. The drying process uses warm air circulated through the cylinder to vaporize any residual solvent. The solvent is recovered and purified in a still. The clean solvent is then pumped back into the holding/base tank. The sequence of operations in dry cleaning is explained in Figure 3.4.

Dry cleaning machines are rated on the basis of the loading of fabric (dry weight) the machine can clean per cycle. Machine sizes vary from very small (10 kg) to large (50 kg) capacity. The length of the cleaning cycle depends upon the types of articles cleaned and the degree of soiling. Some heavily stained garments may go through a stain-removal process

Sorting and inspection ⟶ Spot ⟶ Load machine ⟶ Wash ⟶ Extract ⟶ Dry ⟶ Unload ⟶ Press

Figure 3.4 Sequence of operations in dry cleaning.

prior to cleaning to help in better soil and stain removal. A stain-removal technician will treat specific items just prior to cleaning. A lot of effort goes into the process and there are many skilled technicians involved in caring for the garments.

3.2   Drying equipment

Tumble drying is widely followed in cold climates for drying clothes [148,150]. Figure 3.5 shows the diagram of a tumble-drying machine. In a tumble dryer, the wet garments are tumbled in a cage with forced air circulation. The air temperature can be set as required. The duration of drying depends on the degree of wetness of the garments needed to be dried. Tumble drying can make the garment drier than the ambient atmosphere, which is not feasible by outdoor drying. When a gas flame is used as the heat source for drying, the colour of the garments may fade due to the exposure to gas fumes. Care should be taken when heat-set garments are dried by this method.

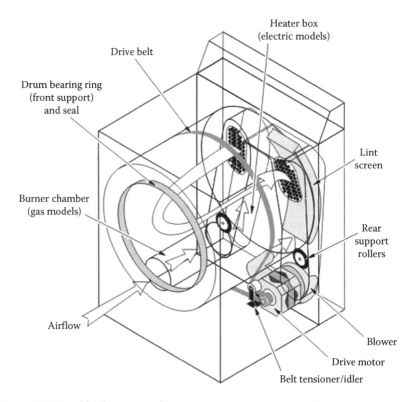

Figure 3.5 Tumble-drying machine.

The tumble-drying machine consists of a large metal drum with paddles around its inner rim. In home machines, the rotation of the drum reverses in every 30 seconds or so, to prevent the bunching up of clothes. In large commercial machines, the drum always rotates in the same direction. Cold air is drawn into the machine through an air intake, often located at the front of the machine. A fan sucks the air in and blows it up towards a heating element. As cool air passes over the heating element, it gets warmed and turned to hot dry air. A thermostat controls the heating element and turns on and off periodically to prevent overheating of the clothes.

Warm air enters the drum through large holes in the back from the heating element. In launderette (large size) machines, the entire drum is full of small holes, and hot air rises up from below. The drum is rotated by a belt connected to the electric motor. Generally, one electric motor drives both the drum and the fan. As the drum rotates, the paddles lift and tumble the wet clothes until they reach the top of the drum. Then the clothes fall back down by the action of gravity. If the dryer is overloaded, the drying load bunches up and rolls around in a big mass instead of tumbling, and hence takes much longer to dry.

The air that leaves the dryer passes through a lint filter that separates dust and fluff. Some dryers are equipped with a second fan to help in extracting the moist air. It is essential to clean the lint filter in a dryer regularly to avoid fire. The air that leaves passes up through a vent hose either mounted permanently in the ceiling or temporarily through an open window. In some dryer designs the humid air exhausted is passed through a heat exchanger and a condenser so the water is cooled and drained away and the heat it contains is captured and reused, increasing the efficiency of the process [149].

On an experimental scale in the industry, infrared (IR) and radio frequency (RF) are also being used to dry garments [274]. However the precise control required for these methods has precluded them from entering the household market.

3.3 Equipment used for pressing

Some of the garments are manufactured as wash and wear or permanent press garments. These clothes are prepared by special finishing treatment to provide crease-resistant properties. For example, the resin treatment of 100% cotton items can provide crease-resistant properties at the expense of loss in strength and abrasion resistance. The manufactured garments can be treated with the specialty chemicals by dipping in the chemical, spraying and vapour-phase treatment. In the first two cases, the garment needs to be cured after the application of the chemicals.

The hand irons are the most common type of pressing equipment used by a household [153]. These irons are heated electrically with the provision of steam supply and temperature control. There are various shapes of the irons and the weight ranges from about 1 kilogram to 15 kilograms. There are several types of pressing tables available for these irons, which may include a simple table or a table with vacuum arrangement to hold the garment or section of a garment in place and dry after pressing. Additional parts can be attached to the table to support various parts of a garment so that a suitable shape is available for each (Figure 3.6).

Steam presses are used to assist in better shape retention and improve the efficiency of pressing. The steam presses can be of various shapes with automatic operations. There are provisions of steam supply to all the parts, vacuum and altering the pressure. Some designs can be fitted with a programmed logic circuit to work in varying cycles depending on the type of garment. In some designs, additional extensions such as bucks or matching heads can be attached when the shape of the garments changes. Many other types of pressing equipment are available that will enhance the final quality of the garments produced. Depending on the type of product being produced, different equipment will be required. Some examples include carousel press, specially designed press for trousers and skirts, steam air finisher and steam tunnel.

Specially designed machines are available for creasing and pleating. Creasing equipment is used to press the edges of clothing components so that they are easily sewn. For example, the cuffs and patch pockets

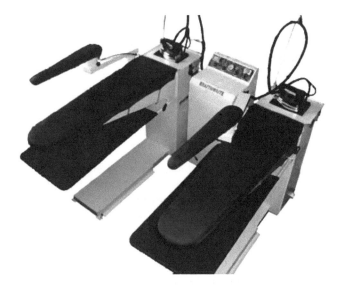

Figure 3.6 Pressing equipment to suit various shapes of a garment.

are formed into shape by the working aid, and are pressed to retain the shape, which makes the sewing operation easier. Pleating machines create a series of creases following a specific pattern or randomly depending on the type of the cloth. Pleats of various lengths can be prepared by hand or by using machines. Blade-type and rotary-type machines can be used for rapid and accurate pleat creation.

Incorrect selection of parameters can lead to shrinkage, colour loss or degradation of the fabric. Hence, all the parameters should be precisely controlled to avoid any damage to the garments. In addition, the accessories used should be able to withstand the processing conditions. In several instances, the lack of understanding of the material and the process can cause permanent damage to the batch of garments. Hence, a perfect understanding and training of the operators is essential to avoid such problems.

3.4   Other equipment

Pressing or ironing is a common method of removing unwanted wrinkles and reinforcing desired creases in garments. The flat iron is a hot plate (heated by coal and later by electricity) that is used for this purpose. The garment to be pressed is spread on a flat surface (ironing board) and then the iron is run over it with a desired amount of pressure. The advent of stainless steel provided a smooth surface that has been enhanced by Teflon coatings. Water, sprinkled by hand on the garment, would turn into steam on contact with the hot iron and aid in removing the creases. Nowadays a water tank is incorporated in the iron to provide precise sprinkling. In advanced systems, this water is converted to steam and then sprayed. Modern irons possess a variety of temperature settings from mild for delicate items to extremely high for linens. The objective is to obtain clean crisp creases.

chapter four

Care labelling

Care labels provide information on the temperature settings to be used for ironing [4]. In certain cases, ironing is not recommended at all. Apparel and textiles are soiled during their normal use. From an economic point of view, these items must be cleaned and refurbished for reuse without substantially altering their functional and aesthetic properties. It is essential that the various processes to which the apparel is subjected to should maintain and restore the desirable and functional properties. This is a joint responsibility of the textile and apparel industry, the textile care industry and the consumers.

The Federal Trade Commission (FTC) in the United States promulgated a trade rule on care labelling of wearing textiles and certain piece goods in 1972 [275]. The rule requires that apparel items should have a permanent care label that provides information about their regular care instructions. The purpose of the rule is to give the consumer accurate care information to extend the useful life of the garments [276]. In addition to the apparel products, the other textile products should also contain information on care labelling. The list of items that use care labels includes clothing, household textiles, piece goods and yarns made from textiles, furnishings, upholstered furniture, bedding, mattresses, bed bases, plastics and plastic-coated fabrics, suede skins, hides, grain leathers and/or furs and custom-made garments (e.g., wedding dresses, suits). However, some of the items do not need care instructions to ensure that the product is not damaged during cleaning and maintenance. The list of items excluded from care labelling includes second-hand goods, footwear, jute products, drapery, haberdashery, some types of furnishings, medical and surgical goods, canvas goods and miscellaneous items such as cords, toys, umbrellas and shoelaces.

Care symbols provide all the necessary information on washing, bleaching, ironing, dry cleaning and tumble drying [277]. The consumer usually does not have the experience or technical knowledge to decide which care treatment is suitable, so it is the responsibility of the apparel manufacturers to provide the necessary care information for the products. All textile used in apparel and all piece goods sold for making home-sewn apparel are covered in care labelling apart from shoes, belts, hats, neckties, non-woven and one-time garments.

Without care information, the consumers will face trouble in deciding on the appropriate conditions for care treatment of the apparel [9]. Care labels should not be considered as a guarantee or a quality mark of the product. The following people or groups are covered in care labelling:

- Manufacturers of textiles and apparel.
- Manufacturers of piece goods sold at a retail price to consumers for making wearing apparel.
- Importers of wearing apparel and piece goods for making wearing apparel.
- Any organization that directs or controls the manufacture and/or import and export of textile wearing apparel or piece goods for making wearing apparel.

The FTC of the United States regulates care labelling for the domestic apparel market under rule 16 CFR Part 423 titled 'Care Labelling of Textile Wearing Apparel and Certain Piece Goods'. The latest amendments state that manufacturers can use a set of four basic care label symbols developed by the American Society for Testing and Materials (ASTM) instead of using words [7]. These symbols are a set of graphic images that function like universal symbols on highway signs that do not need to be translated into a variety of languages. Consequently, products sold in the United States can use text only, symbols only, or both text and symbols. Products that are destined for multiple countries should adopt the symbols-only format.

The information on care labels must be readily understood by consumers in the post-purchase stage [4,10]. Care labels are expected to carry information on fibre type, country of origin, registered identification number (RN), wash-care instructions, size and the manufacturer's or retailer's identification. It becomes the responsibility of the garment manufacturer or retailer to ensure accuracy and validity of the label contents. Consumers with a high need for cognition prefer text format labels while those with a lower need prefer the information in symbol format. Generally, text-based labels are preferred as reading skills are taught and reinforced from an early age. Easily understood labels increase consumer confidence in caring for the apparel and reduce perceptions of risk concerning the purchase of the item.

4.1   Definition of a care label

According to ASTM D 3136-96, a care label is a label or other affixed instructions that report how a product should be. Similarly, care instructions are a series of directions that describe practices that should refurbish a product without adverse effects and warn against any part of the directions that one could reasonably be expected to use that may harm the item [4,278]. The FTC definition states that a "Care label means a permanent label or tag, containing regular care information and instruction, that is attached or affixed in some manner that will not become separated from the product and will remain legible during the useful life of the product".

The care label informs sales personnel and consumers of the appropriate care and treatment of the textile and the other material used in its production. Correct labelling and careful compliance with the information given on the care label help to ensure a long life for the textile items. Care labels help to prevent irreversible damage to the textile article during its care processes [9]. The care labels generally contain the following information, although the statutory provisions may vary from country to country:

- Care symbols
- Fibre content (% of each fibre)
- Size
- Country of origin
- Further information, such as eco labels, etc.

4.2   Terminologies used in care labelling

Besides the above processes used for care labelling, various other terminologies related to care labelling that the consumers should understand thoroughly are described below [2].

1. **Detergent:** A cleaning agent containing one or more surfactants as the active ingredient(s).
2. **Soap:** A cleaning agent usually consisting of sodium or potassium salts of fatty acids.
3. **Bleach** (in care of textiles): A product for brightening and aiding the removal of soils and stains from textile materials by oxidation that is inclusive of both chlorine and non-chlorine products.
4. **Cleaning agent:** A chemical compound or formulation of several compounds that loosens, disperses, dissolves, or emulsifies soil to facilitate removal by mechanical action.
5. **Consumer care:** Cleaning and maintenance procedures as customarily undertaken by the ultimate user.
6. **Professional care:** Cleaning and maintenance procedures requiring the services of a person specially trained or skilled in their use.
7. **Refurbish:** To brighten up or refresh and restore to wearability or use by cleaning such as dry cleaning, laundering or steam cleaning.
8. **Stain removal:** A cleaning procedure for localised areas with cleaning agents and mechanical action specific to the removal of foreign substances present.
9. **Solvent relative humidity:** The humidity of air over dry cleaning in equilibrium with the solvent and a small amount of water.

The care labels for various types of clothing should be positioned at appropriate places in a particular clothing style (see Table 4.1). However, individual manufacturers can slightly vary these positions, but should follow the instructions described in the "Care label requirements" section below.

Table 4.1 Positioning of care labels in various garments [2,279]

Type of garment	Position of label	Type of garment	Position of label
Clothing for men and boys			
Coats, formal jacket, overalls	On the left inside breast pocket; if there is not one, on the lining or the facing, on the left side.	Shirts	At the back or near the neck; if possible on the side seam above the hem.
Sports jacket	On the left inside breast pocket; if there is not one, on the lining or the facing.	Ties	On inside.
Trousers	At the top centre of the right rear pocket; if there is not one, on the waistband at the back.	Pre-packed shorts	At the back on the inside, in the middle of the waistband.
Ski pants and trousers to be worn with a belt (knitted)	On the waistband at the back; if there is not one, at the top of the back centre seam.	Pre-packed vests Swimming trunks	At the top centre back. At the top of the left-side seam.
Clothing for women and girls			
Coats, suits	On the lower front facing. If there is no facing, or if, after making up, it is not suitable for carrying a label, at the top centre back.	Overalls and jumpers	At the top centre back (or, if the material is transparent or the overalls have no neck, in the left-side seam above the hem).
Dresses	When fashion permits, at the top centre back; otherwise on the left-side seam, above the hem.	Pinafore dresses	At the top centre back (with size indication).

(Continued)

Table 4.1 (Continued) Positioning of care labels in various garments [2,279]

Type of garment	Position of label	Type of garment	Position of label
Blouses	At the bottom, on the left-side seam, above the hem.	Underwear	At the top centre back. Exception: For cami-knickers, in the middle of the side seam.
Skirts, trousers	At the waistband, at the top centre back.	Aprons	At the joint between the body of the apron and the left tie.
Corsetry			
Brassieres, short or long	At the back left, at the lower edge of the garment.	Non-stretch corselettes, woven corselettes (non-stretch)	At the back left, at the lower edge of the garment.
Stretch girdles short or long, stretch panty girdles short or long, stretch corselettes	At the top centre back.	Athletic support (non-stretch), suspender belt (non-stretch)	At the back left, at the lower edge of the garment.
Tights	At the centre back.	Stockings	On the packing.
Women's swimwear			
One piece	At the top of the left-side seam.	Two piece Top piece Bottom piece	At the top of the left-side seam. At the top of the left-side seam.
Clothing for men, boys, women and girls			
Pullovers, knitted waistcoats, knitted jackets, anoraks, track suits, tops and bottoms, nightwear for men, women and children	At the top centre back.	Reversible anoraks, dressing gowns, housecoats, bath robes, baby clothes (excluding nappies)	In the pocket. At the neck, beside the size marking. For articles with side seams, on the left one. For articles without side seams, on the left shoulder seam. For all-in-one rompers, on the top outside hem.

(Continued)

Table 4.1 (Continued) Positioning of care labels in various garments [2,279]

Type of garment	Position of label	Type of garment	Position of label
		Other articles	
Table and bed linen (white or coloured) not to be boiled (i.e., easy care articles)	At the corner and on the underside, in the hem.	Shawls and neckerchiefs, finished curtains	At the corner. On the draw tape.
Hand and bath towels, not to be boiled	At the corner or, if possible, on the hanging tab.	Woven and knitted gloves	On the left glove.
Ribbon sold by length, pre-packed	On the packing.	Hand knitting yarn	On the band.
Patterned covers, wool covers (blankets)	At the corner, with the marking.	Hand embroidery yarn, crochet yarn, hand knitting wool	On the band.

4.3   Care label requirements

According to FTC rules, anyone dealing with apparel must establish a suitable basis for care information and it should be sufficient to keep the garment safe during its use [280]. Various care labelling systems are followed worldwide but may differ in terms of the symbols or the wordings that convey the message. Whatever the system may be, it should follow the following set of guidelines [276].

- All the symbols used in the care labelling system should be placed directly on the article or on a label that shall be affixed in a permanent manner to the article.
- The symbols may be produced by weaving, printing or other processes.
- Care labels should be made of suitable material with resistance to the care treatment indicated on the label at least equal to that of the article on which they are placed.
- Label and symbols should be large enough so that they are easily visible and readable.
- All the symbols should be used in the prescribed order and denote the maximum permissible treatment.

- The consumers should easily understand the symbols irrespective of the language.
- The care instruction symbols are applicable to the whole garment including trimmings, zippers, linings, buttons, etc. unless otherwise mentioned by separate labels.
- The care symbols selected should give instructions for the most severe process or treatment the garment can withstand while being maintained in a serviceable condition without causing a significant loss of its properties.
- The label, with the symbols and words on it, should be legible throughout the useful life of the garment.
- The machines used for washing and drying should be able to provide the conditions mentioned on the care label.
- The care labels should not be visible from the outside of the garment.
- They should not be inconvenient or cause irritations to the wearer.
- They should be easily visible and not hidden, which would otherwise lead to difficulties in conveying information.
- The labels for a particular style should be positioned at one place, either on the back, top or middle.
- If not readily seen due to packaging, care information must be repeated on the outside of the package or on a hangtag attached to the product.
- It is not always possible to have all the information on one label due to the type of the garment, material and fashion requirements. In these cases it is permissible to go for a second label.
- When a garment consists of two or more parts and is always sold as a unit, only one care label can be used if the care instructions are the same for all the pieces. The label should be attached to the major piece of the suit. If the suit pieces require different care instructions or are designed to be sold separately, like coordinates, then each item must have its own care label.
- When an article is made of different materials, care instructions suitable for all materials, including those most sensitive should be included.
- The textile items that can be neither laundered nor dry cleaned, should indicate this on the care label and adequately describe the recommended care treatment.
- Sometimes delicate components of articles such as trims or padding in furniture or bedding cannot or should not be removed. In these cases, suppliers should also consider the care of the delicate parts when developing care instructions for the article.

The care labels should be used for a wide range of products such as apparel textiles, household textiles, home furnishings, resin-coated fabrics, piece

goods made from textiles, suede skins, leathers and furs [9,13,281]. Care labelling used in clothing should provide the consumers enough information on
(1) care instructions for clothing and other textile products, (2) prior knowledge of care and maintenance costs of the materials such as dry cleaning, (3)
the processes and conditions to avoid in order to maximize the useful life of
clothing and textile products and (4) possible damages that can happen such
as dyes running out (e.g., wash separately) during care and maintenance.

The information provided on the care labels also affects the purchase
decision of the consumers [276,282–284]. During clothing purchase, consumers are seeking the information on the fibre content [285] and the cost involved
with the care procedures [285] in addition to the price, physical characteristics (colour, size and style) and brand name. A permanent label containing
care instructions should comply with the following requirements:

1. The label shall be accessible for examination by a prospective consumer as an integral part or on a removable label attached to the
   article.
2. The labels may use easily understood symbols in addition to written instructions. Symbols and letters on labels must remain legible
   throughout the useful life of a garment. Medium-width lettering
   of which no individual letter shall be less than 1.5 mm high should
   preferably be used.
3. Each significant component should possess its own appropriate and
   complete care instructions.
4. The care instruction symbols are applicable to the whole of the garment, including trimmings, zippers, linings, buttons and sewing
   thread, unless otherwise mentioned on separate labels.
5. Labels for a particular style should be positioned at one place in all
   garment pieces.
6. Care labels should be made of suitable material with resistance to
   the care treatment indicated in the label at least equal to that of the
   article on which they are placed.
7. The symbols selected should give instructions for the most severe
   maintenance process or treatment the garment can withstand without causing a significant loss of its properties.
8. Care instructions are chosen based on the end-use of the article and
   fibre type. In case of a fabric with blended fibres, the care instructions
   should generally be based on the properties of the most sensitive fibre.
9. The care instructions selected should be verified to ensure that the
   article complies with the performance requirements prior to sale.

Care instructions may be woven into or printed on labels primarily of a
rectangular shape [4,281]. They may be loop labels (sewn at both ends) or
fused flat against the fabric without the use of sewing threads. The size

of the labels will depend on the amount of information with an emphasis on legibility. The position of the label is not stipulated and hence may be placed at suitable locations in different garments. Although the rule states that the organisation or person controlling the manufacture of the finished garment is responsible for care labelling, the burden of proof falls on the consumer if an item fails to meet the stated performance standards.

The suppliers are legally responsible to ensure that the clothing and textile products they supply should satisfy the mandatory requirements. Legal action or penalties apply when they fail to comply. Hence, they can avoid the penalties by the following steps:

- Understand the specific requirements for a textile item.
- Understand the Consumer Protection Act and regulations.
- They should ensure that the products comply with the requirements of this mandatory standard by thoroughly testing the products.

Similarly, retailers should ensure that the products they supply should satisfy the mandatory standards. The retailers should always:

- Stipulate that any textile item they order must meet the mandatory standard.
- Undertake visual inspection of delivered stock to check compliance with the requirements of the mandatory standard.
- Encourage consumers to understand and follow the care instructions to avoid damage and maximise the useful life of clothing and textile products they supply.

Some important difficulties with care labels are (1) some indicate highly restrictive procedures that may not really be necessary, (2) some instructions are difficult to understand and (3) some abrasive and coarse labels cause skin irritation. A survey confirmed that many people use inappropriate cleaning methods because they do not correctly understand care label information. Some respondents indicated that they thought bleaching was acceptable, though the instruction warned against it. Similarly, 'line dry' was interpreted incorrectly. Educational programmes are therefore necessary to enlighten the consumers. Standardising information on care labels can also minimise misunderstanding. Considerable evidence also exists to demonstrate that there is no direct relationship between information provision on care labels and information used.

4.4   Mandatory regulations

In various countries there are mandatory requirements for care labelling. Table 4.2 describes the standards used in various countries for care labelling.

Table 4.2 Standards for care labelling

Standard	Description
AS/NZS 1957:1998	Textiles: care labeling.
AS/NZS 2622:1996	Textile products: fibre content labeling.
ASTM D5489-14	Standard guide for care symbols for care instructions on textile products.
ASTM D3136-14	Standard terminology relating to care labelling for apparel, textile, home furnishing and leather products.
ASTM F2061-12	Standard practice for chemical protective clothing: wearing, care and maintenance instructions.
ASTM F1449-08	Standard guide for industrial laundering of flame-, thermal- and arc-resistant clothing.
ISO/TR 2801:2007	Clothing for protection against heat and flame – general recommendations for the selection, care and use of protective clothing.
ISO/TR 21808:2009	Guidance on the selection, use, care and maintenance of PPE designed to provide protection for firefighters.
ISO 3758:2012	Textiles: care labelling code using symbols.
ISO 3175-1:2010	Textiles: professional care, dry cleaning and wet cleaning of fabrics and garments-Part 1: assessment of performance after cleaning and finishing.
ISO 3175-2: 2010	Textiles: professional care, dry cleaning and wet cleaning of fabrics and garments-Part 2: procedure for testing performance when cleaning and finishing using tetrachloroethene.
ISO 3175-3:2003	Textiles: professional care, dry cleaning and wet cleaning of fabrics and garments-Part 3: procedure for testing performance when cleaning and finishing using hydrocarbon solvents.
ISO 3175-4:2003	Textiles: professional care, dry cleaning and wet cleaning of fabrics and garments-Part 4: procedure for testing performance when cleaning and finishing using simulated wet cleaning.
ISO 6330: 2012	Textiles: domestic washing and drying procedures for textile testing.
ASTM E2274 and E2406	Standard test method for the evaluation of laundry disinfectants and sanitizers.
NFPA 1851	Standard on selection, care and maintenance of protective ensembles for structural firefighting and proximity fire fighting.

(Continued)

Table 4.2 (Continued) Standards for care labelling

Standard	Description
NFPA 2113	Standard on selection, care, use and maintenance of flame-resistant garments for the protection of industrial personnel against short-duration thermal exposures.
AS/NZS 4501.1:2008	Occupational protective clothing – guidelines on the selection, use, care and maintenance of protective clothing.
AS/NZS 2161.1:2000	Occupational protective gloves – selection, use and maintenance.
AS/NZS ISO 2801	Clothing for protection against heat and flame – general recommendations for selection, care and use of protective clothing.
ISO 15797:2002	Textiles: industrial washing and finishing procedures for the testing of work wear.
ISO 30023:2010	Textiles: qualification symbols for labelling work wear to be industrially laundered.
AS/NZS 2621:1998	Textiles: guide to the selection of correct care labelling instructions from AS/NZS 1957.
AS/NZS 2622:1996	Textile products: fibre content labeling.

The International Organization for Standardization (ISO) is currently preparing a system of care labelling with special symbols for the industrial cleaning of work wear. This system of graphical symbols is intended for the care labelling of work wear articles and protective clothing. This standard will provide information on the professional industrial laundering using ISO 15797, as the industrial cleaning processes (in ISO 15797) are fundamentally different from the GINETEX symbols (ISO 3758) used on retail articles intended for home laundering. GINETEX symbols are not applicable to professionally cleaning the garments. Industrial laundering treatments such as washing, bleaching, tunnel finishing and tumble drying after washing are covered in this standard. ISO 30023 does not give care instructions via the label. Instead, it only shows that the textiles have been tested for resistance/suitability to the industrial laundry process (as defined in ISO 15797) as shown in Table 4.3.

ISO 30023 does not use a label to give care instructions. It only shows that the textiles have been tested for resistance/suitability to the industrial laundry process (that is defined in ISO 15797). The scope of ISO 30023 is limited due to the scope of ISO 15797 (to work wear that can be washed and dried using the 8+2 procedures).Testing the samples as per ISO 15797 is expensive as large-scale machineries and more tests are involved in this standard (Figure 4.1).

Table 4.3 ISO 30023 (new symbols)

Symbol	Explanation
PRO ▭ ⬡	Professional industrial laundering, resistant to ISO 15797 washing procedure 1, tunnel finish or tumble dry.
PRO ⬡	Professional industrial laundering, resistant to ISO 15797 washing procedure 3, tumble dry only.
PRO ▤	Professional industrial laundering, resistant to ISO 15797 washing procedures 1 and 6, tunnel finish only.
⬡	Tumble drying.
▤	Tunnel/cabinet finishing/drying.

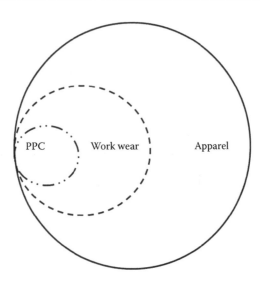

Figure 4.1 The scope of ISO 15797 and 30023.

There are certain advantages of ISO 30023 as mentioned below:

- By using the symbols in ISO 30023, manufacturers can communicate easily on the resistance of their textiles to industrial laundry (using ISO 15797 reference procedures).
- Symbols will provide a picture of results of ISO 15797 – testing for resistance to industrial laundry.
- Unlike domestic care labelling systems (CLS), manufacturers remain free on care and maintenance instructions.
- ISO 30023 will facilitate the exchange of technical information along the supply chain.
- The cost of testing, using ISO 15797, will be considerably lower than the savings it will generate.
- ISO 30023 will speed up the delivery to end-users of textiles suitable for industrial laundry.
- ISO 30023 will lead to improved accuracy in specifications for the supply of textile items.
- Linking labelling with recognized test methods will ensure product quality and lead to less disputes.

4.5   Processes described by care labels

Care labels describe useful information on the processes used for care and maintenance of clothing items, which include laundering or washing, bleaching, ironing, dry cleaning and tumble drying (Table 4.4).

Table 4.4 Basic symbols used for care instructions in care labels

Symbols	Indicates
	Wash tub: gives instructions about laundering.
	Triangle: gives instructions for bleaching.
	Square: relates to drying.
	Hand iron: provides ironing or pressing instructions.
	Circle: gives dry-cleaning instructions.

4.5.1   Laundering

A process intended to remove soil or stains by treatment (washing) with an aqueous detergent solution (and possibly bleach) and normally including subsequent rinsing, extracting and drying. The process may be further divided as hand washing, home laundering and commercial laundering.

- **Hand washing:** The gentlest form of home laundering using hand manipulation without the use of a machine or device such as a scrubbing board.
- **Home laundering:** A process by which textile products or parts thereof may be washed, bleached, dried and pressed by non-professionals.
- **Commercial laundering:** A process by which textile products or specimens may be washed, bleached, rinsed, dried and pressed typically at higher temperatures, higher pH, and longer times than used for home laundering.

The process may include various operations in relevant combinations such as soaking, pre-washing and proper washing carried out usually with heating, mechanical action and in the presence of detergents or other products and rinsing. Water extraction, i.e., spinning or wringing, is performed during and/or at the end of the operations mentioned above.

4.5.2   Bleaching

Bleaching helps to remove stains on white clothes and retain their brightness. It can remove the colour when used on coloured clothes. The bleaching agents can be classified as (1) chlorine bleach and (2) non-chlorine bleach.

1. **Chlorine bleach:** A process carried out in an aqueous medium before, during or after washing processes, requiring the use of a chlorine-based bleaching agent for the purpose of removing stains and/or improving whiteness.
2. **Non-chlorine bleach:** Bleach that does not release the hypochlorite ion in solution, i.e., sodium perborate, sodium percarbonate, etc.

4.5.3   Dry cleaning

Dry cleaning is the process of cleaning textile articles by means of organic solvents (e.g., petroleum, perchlorethylene and fluorocarbon). This process consists of cleaning, rinsing, spinning and drying.

4.5.4 Tumble drying

A process carried out on a textile article after washing with the intention of removing residual water by treatment with hot air in a rotating drum.

4.5.5 Ironing or pressing

Ironing is a method that uses a heated iron (with or without the presence of steam) to smooth or retain the shape of a garment by the application of heat and pressure.

4.6 Care labelling systems

At present, there is no universal care labelling system. In the United States, the Wool Products Labelling Act (1938), the Fur Products Labelling Act (1951), the Flammable Fabrics Act (1958) and the Rule on Care Labelling (1972) are in force. The Japan Industrial Standards (JIS) for care labelling came into force in 1962. Similarly, in Korea, the rule on Quality Labelling came into force in 1969 and the use of symbols for care labelling of apparel products was published in 1972.

The ASTM system is accepted in the North American Free Trade Agreement (NAFTA) countries. The ISO or the International Association for Textile Care Labelling (GINETEX) system is accepted in most of Europe and Asia. Japan has its own system. Negotiations are under way to harmonise the two major systems (ASTM and GINETEX) into a universal labelling system for care procedures. An international labelling system can facilitate global trade by avoiding technical or standards barriers. The major systems that are followed worldwide are ASTM, ISO (GINETEX) and the Canadian, Dutch and Japanese Care Labelling Systems. Although there are some variations in the symbols, the five basic symbols used in many of the above listed systems are discussed in a Section 4.6.1.

The Canadian Care Labelling System employs five basic symbols that are sub-divided into three conventional traffic light colours [4]. The red (with a cross superimposed) indicates prohibition, yellow the need of care and green indicates that no special precautions need be taken.

The JIS for care labelling came into force in 1962. Similarly, in Korea, the rule on Quality Labelling came into force in 1969 and the use of symbols for care labelling of apparel products was published in 1972.

The Technical Committee (TC-38) of ISO handles all types of textile standards through several subcommittees. Subcommittee SC-11 is concerned with developing standards for care labelling with the primary objective of developing an international symbol system. Manufacturers and retailers follow the ASTM standard (ASTM D 3938 – determining or confirming care instructions for apparel and other textile consumer products) to ensure correct information is included on care labels. The other standards dealing with care labelling include ASTM D 3136 (standard terminology relating to care labels for textile and leather products other than textile floor coverings and upholstery), ASTM D 6322 (international test methods associated with textile care procedures) and ASTM D 5489 (standard guide for care symbols for care instructions).

The most recent amendment to the rule states that the manufacturers can use a set of basic care label symbols developed by ASTM instead of using words. These symbols are graphic images that function like universal symbols on highway signs and that do not need to be translated into a variety of languages. The intention of using symbols is that an individual without any previous experience should be able to interpret the symbols correctly and follow the actions suggested by them. However, it is often hard for the consumers to understand the symbols correctly. Following the FTC rules, products sold in the United States can use text only, symbols only or both text and symbols. Products that are destined for multiple countries should adopt the symbols-only format to avoid the need to label in multiple languages. Consumers with a high need for cognition prefer labels that present care information in text format, while those with a lower need prefer the information in symbol format.

Care labels that are easily understood by consumers increase their confidence in caring for the apparel and reduce their perceptions of risk concerning the purchase of the item. The care instructions can be passed to the consumers with text only, symbols only and a combination of text and symbol [286]. The manufacturers of clothing and other textiles prefer to use symbols on the care label, as the care symbols are globally recognizable and do not need to be translated into other languages. The majority of the consumers prefer care labels that contain text and symbols, as these skills are taught and reinforced from an early age.

There are various care labelling systems followed around the world. Among the different care labelling systems, the systems that are mainly followed are listed below:

- International Care Labelling System
- ASTM Care Labelling System
- Canadian Care Labelling System
- British Care Labelling System
- Japanese Care Labelling System
- Australian Care Labelling System

4.6.1   International (ISO) care labelling system

The ISO system commonly known as GINETEX for care labelling was established in 1963 in Paris following several international symposiums for Textile Care Labelling at the end of the 1950s [287,288]. A large number of national organisations are members of GINETEX. The countries that are members of GINETEX are Austria, Brazil, the Czech Republic, Denmark, Finland, France, Belgium, England, Germany, Greece, Italy, Lithuania, the Netherlands, Portugal, Slovakia, Slovenia, Spain, Switzerland, Tunisia, Turkey and the United Kingdom. GINETEX has the following objectives:

- To define symbols for textile care at an international level.
- To define the regulations for the use of care symbols.
- To promote the use of clothing care symbols.
- To acquire all markings and all rights relative to the symbols.
- To register the symbols, both national and international.
- To insure protection for all marks and symbols as adopted in all member countries of GINETEX.
- To conclude all agreements liable to the promotion of the above-mentioned objectives
- To take all measures and carry out all actions in order to promote the above objectives, either directly or indirectly.

An internationally applicable care labelling system based on symbols for textile materials has been devised by GINETEX. The care labelling system provides the correct information on the care instruction of textile products to consumers, retailers and textile manufacturing companies. The care labels describe various processes the clothing item can tolerate to avoid any irreversible damage to the product. The symbols or pictograms used in most countries are registered trademarks of GINETEX.

The international system of care labelling symbols is defined by GINETEX. In addition, GINETEX promotes and coordinates its technical background on an international level. The care labelling system takes into account any new technical and ecological developments together with changes in consumer practices. The symbols used in the GINETEX system represent that the garment can withstand the process and a cross indicates the process is not possible for the garment. The five symbols as described in Figure 4.2 are used in this system.

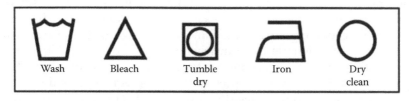

Wash   Bleach   Tumble   Iron   Dry
                          dry   clean

Figure 4.2 Five symbols used in the ISO system for care labelling.

The washtub represents the washing or laundering process, which may contain some number inside. This number indicates the maximum permissible temperature of the water in degrees centigrade. Both the washtub and the number indicate that machine washing is possible. A hand in the washtub indicates only hand washing is possible. If there is an underline beneath the washtub, it indicates a milder treatment is in order. Numbers above the washtub indicate different washing programs and these are not always identical with those actually used in washing machines. There may be some additional indications that are not followed everywhere. 'CL' inside the triangle indicates that chlorine bleaching is possible. The dots (1, 2 or 3) inside the iron symbol indicate the maximum temperature at which ironing can be done. The letters (A, P or F) inside the circle indicate the dry cleaning process with the solvent to be used (A, P and F indicate any solvent, any solvent except trichloroethylene and petroleum solvent, respectively). A circle inside a square indicates that a particular garment can be tumble dried.

4.6.2 ASTM care labelling system

In the ASTM system there are five basic symbols: washtub, triangle, square, iron and circle indicating the process of washing, bleaching, drying, ironing or pressing and dry-cleaning, respectively [7]. The prohibitive symbol 'X' may be used only when evidence can be provided that the care procedure on which it is superimposed would adversely change the dimensions, hand, appearance or performance of the textile. The washtub with a water wave represents the washing process in the home laundering or commercial laundering process. Additional symbols inside the washtub represent the washing temperature and the hand washing process. The water temperature in the hand washing process may be 40°C. Additional symbols below the washtub indicate the permanent press cycle (one underline, minus sign, or bar) and a delicate-gentle washing cycle (two underline, minus sign, or bar). The detailed description is given in Figure 4.3.

The triangle represents the bleaching process and an additional symbol inside the triangle indicates the type(s) of bleach to be used. The square indicates the drying process. An additional symbol inside the square represents the type of drying process to use including tumble dry, line dry, drip dry, dry flat and dry in shade. Additional symbols below the square indicate the permanent press cycle (one underline, minus sign or bar) and the delicate-gentle cycle (two underline, minus sign or bar). Permanent press and gentle/delicate cycle instructions may be reported in words along with symbols instructions for tumble-drying and the dryer heat setting. The dots are used to represent the dryer temperature: three dots (high), two dots (medium), one dot (low), no dots (any heat) and a solid circle (no heat/air).

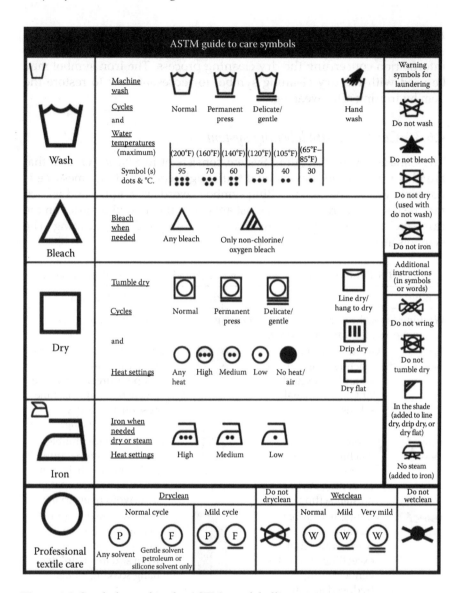

Figure 4.3 Symbols used in the ASTM care labelling system.

The hand iron represents the hand ironing process and the pressing process. Dots inside the iron represent the temperature setting. The maximum temperatures by different number of dots are three dots (200°C), two dots (150°C) and one dot (110°C). The warning symbol, crossed-out steam lines under the iron or words may be used to report the warning, 'do not steam'. The circle represents the dry cleaning process.

A letter inside the circle represents the type of solvent that can be used. Additional symbols may be used with the circle to furnish additional information concerning the dry cleaning process. The iron symbol may be used with the dry cleaning symbol to represent how to restore the item by ironing after wearing.

4.6.3 Canadian care labelling system

The Canadian Care Labelling System consists of five basic symbols that are illustrated in the conventional traffic light colours. If any message is not conveyed by the care labelling symbols, words in English and French may be used. The five symbols must appear in the following order on the care labels: washing, bleaching, drying, ironing and dry cleaning. The symbols are described in Table 4.5.

Table 4.5 Symbols and processes used in Canadian care labelling system

Symbol	Instructions	Symbol	Instructions
Washing (a washtub without water wave represents washing process)		**Drying** (square symbol indicates drying)	
70°C	Green washtub: machine wash in hot water (not exceeding 70°C) at a normal setting.		Green square: tumble dry at medium to high temperature and remove article from the machine as soon as it is dry. Avoid over drying.
50°C	Green washtub: machine wash in warm water (not exceeding 50°C) at a normal setting.	◯	Orange square: tumble dry at low temperature and remove article from machine as soon as it is dry. Avoid over drying.
50°C	Orange washtub: machine wash in warm water (not exceeding 50°C) at a gentle setting (reduced agitation).		Green square: hang to dry after removing excess water. Green square: 'drip' dry, hang soaking wet.
40°C	Orange washtub: machine wash in lukewarm water (not exceeding 40°C) at a gentle setting (reduced agitation).		Orange square: dry on flat surface after extracting excess water.

(Continued)

Table 4.5 (Continued) Symbols and processes used in Canadian care labelling system

Symbol	Instructions	Symbol	Instructions
	Orange washtub: hand wash gently in lukewarm water (not exceeding 40°C).	Ironing (the iron with a closed handle represents the ironing process).	
	Yellow washtub: hand wash gently in cool water (not exceeding 30°C).	and	Green iron: iron at a high temperature (not exceeding 200°C – recommended for cotton and linen).
	Red washtub: do not wash.	and	Orange iron: iron at a medium temperature (not exceeding 150°C – recommended for nylon and polyester).
Bleaching (triangle symbol indicates bleaching process)		and	Orange iron: iron at a low temperature (not exceeding 110°C – recommended for acrylic).
	Orange triangle: use chlorine bleach with care. Follow package directions.		Red iron: do not iron or press.
	Red triangle: do not use chlorine bleach.	Dry-cleaning (the circle indicates dry-cleaning)	
			Green circle: dry clean Orange circle: dry clean, tumble at a low safe temperature. Red circle: do not dry clean.

4.6.4 British care labelling system

The British Care Labelling System uses graphic symbols to provide information on care labels. The care instruction symbols should appear in the order of washing, bleaching, drying, ironing and dry cleaning. The symbols used for various processes with necessary explanation are given in Table 4.6.

Table 4.6 Symbols and processes used in British care labelling system

Washing

	Cotton wash (no bar)	A washtub without a bar indicates that normal (maximum) washing conditions may be used at the appropriate temperature.
	Synthetics wash (single bar)	A single bar beneath the washtub indicates reduced (medium) washing conditions at the appropriate temperature.
	Wool wash (double underline)	A double underline beneath the washtub indicates much reduced (minimum) washing conditions, and is designed specifically for machine-washable wool products.
	Hand wash only	Wash by hand.

Bleaching

Any bleach allowed.

Only oxygen bleach/non-chlorine bleach allowed.

Do not bleach.

Ironing

Hot iron

Warm iron

Cool iron

Dry cleaning

Must be professionally dry cleaned. The letters contained within the circle and/or a bar beneath the circle will indicate the solvent and the process to be used by the dry cleaner.

Must be professionally dry cleaned. The letters contained within the circle and/or a bar beneath the circle will indicate the solvent and the process to be used by the dry cleaner.

(Continued)

Table 4.6 (Continued) Symbols and processes used in British care
labelling system

(W)	Professional wet clean only.
⊠	Do not dry clean.
	Tumble drying
◯	May be tumble dried
⊙⊙	with high heat setting
⊙	with low heat setting
⊠	Do not tumble dry.
✕	A cross through any symbol means 'DO NOT'

Source: http://www.care-labelling.co.uk/whatsymbolsmean.html.

4.6.5   *Australian/New Zealand care labelling system*

The joint systems AS/NZS 1957:1998, Textiles-Care labelling and AS/NZS
2622:1996 Textile products – Fibre content labelling; AS/NZS 1957:1998,
Textiles-Care labelling with variations and additions are followed in
Australia and New Zealand. The joint standards specify that care instructions must be permanently attached to articles, written in English and
appropriate for the care of the article. The instructions can be clarified by
additional symbols if needed, but symbols alone are not sufficient. The
symbols are used for washing, bleaching, drying, ironing and dry cleaning instructions. In addition, professional wet cleaning is explained by the
use of one symbol.

The washtub symbol indicates the washing process. The number
inside the washtub indicates the maximum temperature for washing. A
single line underneath the washtub indicates mild treatment or washing
cycle using the permanent press setting, whereas two lines underneath
the washtub indicate gentle or delicate cycle. A cross indicates that the
process is not appropriate for the clothing.

The bleaching, drying, ironing, dry cleaning and professional wet
cleaning processes are indicated by the symbols as shown in Table 4.7. The
iron symbol indicates ironing and the dots inside indicate the acceptable

Table 4.7 Care symbols used in joint Australian/New Zealand system

Washing

Normal	Permanent press	Delicate/gentle	Hand wash
HOT ●●● 95°C	HOT ●●● 70°C	HOT ●●● 60°C	HOT ●● 50°C
WARM ●● 40°C	COOL ● 30°C	Do not wash	Do not wring

Bleaching

Any bleach (when needed)	Only non-chlorine bleach (when needed)	Do not bleach

Ironing

Cool iron	Warm iron	Hot iron	Do not iron

No steam (added to iron)

Dry cleaning

P Any solvent except trichloroethylene	F Petroleum solvent, liquid silicone	A Any solvent	P Dry clean in accordance with P*

F Dry-clean in accordance with F*	Do not dry clean

* With strict limitation on amount of water, reduced mechanical action and low drying temperature

(Continued)

Table 4.7 (Continued) Care symbols used in joint Australian/New Zealand system

Drying: Tumble drying

Symbol	Normal	Permanent press	Delicate/gentle
	Normal	Permanent press	Delicate/gentle

Symbol	Any heat	High	Medium	Low
	Any heat	High	Medium	Low

No heat/air

Air drying (no tumble dry)

Symbol	Line dry – hang to dry	Drip dry	Dry flat	In the shade
	Line dry – hang to dry	Drip dry	Dry flat	In the shade

Do not dry	Do not tumble dry

Professional wet cleaning

Symbol	Normal process	Mild process	Very mild process	Do not professionally wet clean
	Normal process	Mild process	Very mild process	Do not professionally wet clean

temperature range. The dry cleaning process is indicated by a circle and the letters within the circle provide information on the solvents to be used during the cleaning process, which is needed by professional textile cleaners. The line below the circle indicates the limitations in the dry cleaning process, which may be related to mechanical action, addition of moisture and/or drying temperature.

4.6.6   Japanese care labelling system

The Japanese Care Labelling System uses basic symbols that are different than other systems for care labelling. The symbols are shown in Table 4.8.

Table 4.8 Symbols and processes used in Japanese care labelling system

Symbol	Instructions	Symbol	Instructions
Washing instructions		**Dry cleaning instructions**	
	Machine wash in water temperature of 95°C or less. No other restrictions.		Dry clean; use any dry cleaning agent.
	Machine wash in water temperature of 60°C or less. No other restrictions.		Dry clean; use only a petroleum-based agent.
	Machine wash in water temperature of 40°C or less. No other restrictions.		Do not dry clean.
	Machine wash at delicate cycle or hand wash in water temperature of 40°C or less.	**Wringing instructions**	
	Machine wash at delicate cycle or hand wash in water temperature of 30°C or less.		Wring softly by hand or spin dry by machine quickly.
	Hand wash in water temperature of 30°C or less.		Do not wring by hand.
	Do not wash (not washable).		

(Continued)

Table 4.8 (Continued) Symbols and processes used in Japanese care
labelling system

Symbol	Instructions	Symbol	Instructions
Bleaching instructions		**Drying instructions**	
	Use chlorine bleach.		Hang dry.
	Do not use chlorine bleach.		Hang dry in shade.
Ironing instructions			Lay flat to dry.
	May be ironed directly at 180°C–210°C.		Lay flat to dry in shade.
	May be ironed directly at 140°C–160°C.		
	May be ironed directly at 80°C–120°C.		
	Do not iron.		
	May be ironed at 180°C–210°C if a cloth is placed between the iron and the garment.		

4.7   Example of care labels

An example of a care label is shown in Figure 4.4. This care label explains
the care instructions, type of fibre used, and country of origin, explained
by text as well as symbols. There may be variations to these labels depending on the type of the cloth and country of final destination.

4.8   Electronic care labels

The electronic labels use radio frequency identification (RFID) tags for
storing information electronically on a garment [279,289–291]. The same
RFID tag used for containing the product details at the point of sale cannot
be used for storing the care instructions due to privacy concerns [292,293].
Hence, additional tags are needed for the care instructions, which will
increase the cost of the garment. In addition, the consumers need special
readers to extract the information stored on the RFID tag. Hence, they will
prefer the physical label instead of an electronic label. Furthermore, the

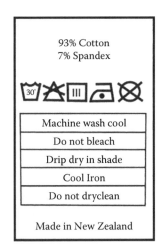

93% Cotton
7% Spandex

| Machine wash cool |
| Do not bleach |
| Drip dry in shade |
| Cool Iron |
| Do not dryclean |

Made in New Zealand

Figure 4.4 Examples of care labels explaining the meanings of various symbols.

use of RFID tags in the garment may pose health risks to the consumers due to the exposure to radiation.

The other concern related to the use of RFID tags is the electronic waste [293]. An appropriate method should be devised to recycle or reuse the RFID tags if they are removed at the point of sale. Similarly, if customers remove the chip at some point, proper disposal of the tags is essential, which causes additional worries to the companies. In addition, the other issues related to RFID are the lack of standardisation and high cost. Standardising the technology, design and use of the tags can solve the issues related to a lack of standardisation, whereas the technological developments can help to overcome the cost-related issues.

RFID tags can be used in combination with the RN system, which is currently in use in some places including the United States. The FTC issues the RN, which can be used on the textile label instead of the company name. The use of the RN system assists the buyers to easily identify a company from the RN using the Internet. The RN system uses less space on the label and more space is available for other care instructions. Therefore, the combined use of RFID and RN number systems could further benefit industry and consumers.

The use of RFID tags is still in its infancy stage [289]. Hence, it is too early to conclude the final shape of RFID tags and their impact on consumers. One of the industry-observed benefits to the consumers is by including washing instructions in the tag. In Germany, RFID chips are being used in nursing homes to help the nurses in correctly sorting the garments according to care instructions. Hence, there is a potential benefit of RFID tags to the consumers. However, all the major drawbacks should be overcome to make RFID tags a success.

4.9   Issues related to care labelling

The selection of processes for garment care and maintenance can increase the cost as shown in Figure 4.5 as well as affect the useful properties. Garments may lose the useful properties due to the deviation from the label instructions and improper conditions used for the care and maintenance [294]. They may lose their visual and tactile aspects or shrink more than the specified values [137]. They may also fade in colour or cross stain during the washing or dry cleaning and subsequent processes. All these consequences result in the rejection of the garment. Hence, consumers should strictly follow the care instructions, otherwise they are solely responsible for the damage caused to the garment.

In some countries, customers demand the use of their national language for the care labelling and marking of textile products. Hence, the use of multi-lingual care label systems exists in some places around the globe [11]. In this case, the labels would have to be very big to accommodate several languages, and the extra cost for industry is mostly associated with translating and relabelling. These problems can be solved by the use of care label instructions in symbols rather than text so that they can be easily followed by all the consumers.

The use of chemical substances is regulated through the European Community Regulation known as the Registration, Evaluation, Authorisation and Restriction of Chemical (REACH) substances, EC1907/2006). However, many consumer organisations around the globe have mentioned that the use of chemical substances in textiles is not sufficiently addressed in the current legislation. The use of nanotechnologies and the residual chemicals present in the garments are neither covered by REACH nor well-documented. Chemical labelling will incur additional

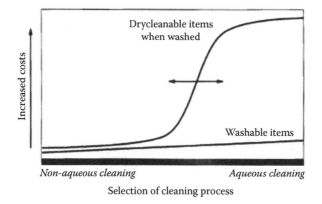

Figure 4.5 Cleaning process and cost.

costs for the industry to conduct tests for chemical substances. Hence, many industries are not in favour of the labelling of chemicals, and they think that labelling will hardly add any value for the consumer.

The residual solvent that remains after dry cleaning can affect the water repellency, reduce the breathability and other properties. Hydrocarbons should be used for olefins, PU and rubber items instead of perchloroethylene (PCE, also called perc) as it can cause swelling to textile items. Elastane fibres swell in PCE and return to the original size after solvent evaporation. Olefin piping in corded seams can shrink and the coating may be separated by PCE.

chapter five

Care instructions for specialty textile items

All textile materials, irrespective of their area of use, get soiled by general use and exposure to the environment, consequently requiring cleaning and care [4]. It is to be noted that, in the house, the use of textiles extends far beyond that of apparel into carpets and rugs, curtains and upholstery material, bed linens and other specialty textiles as in blinds, shutters, awnings, blankets and leather goods. In addition, other sectors such as automotive (for seats and interiors), offices (for separators), hospitality and medical (bed linens) and commercial seating also use textiles.

The manufacturers of carpets and rugs generally give functional finishes that reduce the degree of soiling of the product [295–297]. In addition, they are shampooed and steam cleaned to restore them. Special chemicals (powders or liquids) may also be employed to remove stains. The common issues and cause for carpet replacement are wearing out of the materials, colour fading and fashion. A stain-resistant finish is commonly mandated for curtains and upholstery by the manufacturers. These products are also expected to be resistant to microbial agents, mildew and rot [41]. Vacuum cleaners are widely employed to maintain these materials. Curtains may also be subjected to washing. Replacement is prompted mainly by colour fading on prolonged exposure to sunlight and/or gas fumes (if heating is involved).

Bed linens are washed using harsher settings in a commercial laundry as compared to apparel. This is prompted by the assured cleanliness, critical in the hospitality sector, despite the shortened useful life. Home linens are, however, included in the general wash and undergo a relatively mild treatment. In developed countries, hospital linen has morphed to the disposable kind and after-care is not significant.

Seats in personal automobiles are regularly vacuumed, shampooed and dried. The use of changeable car seat covers is also prevalent. Fabrics used in commercial seating and office separators are dusted or vacuumed, and usually replaced before any significant physical change/damage becomes readily apparent. In the medical field, for disposable applications, products are designed to possess a minimum usage life and appropriate shelf life. For all the above-mentioned items, care instructions are necessary for optimal washing and storage conditions wherein the product will retain its special characteristics.

5.1   Carpet and rugs

Professional carpet care and cleaning is required to maintain the carpets rather than regular vacuuming. However, frequent vacuuming can help to keep grit from becoming embedded into the carpet [298,299]. Most carpets need hot-water extraction, either in isolation or in combination with cleaning, for their maintenance [300]. Some carpet designs may need special means to be cleaned. The carpets should be cleaned by carpet care professionals at least every 18–24 months to refresh the texture and rejuvenate the fibres in the carpet.

Fibre content is the most important parameter in selecting a care process for carpets [101]. Most recent carpets are fabricated from synthetic fibres such as nylon, polyester or polyolefin, which may be cleaned with most cleaning methods. The carpets manufactured from natural fibres such as wool, cotton, silk and sisal require specialised care.

Any attempt to remove the stain by rubbing can lead the spill of the stain into the pile and damage the fibres of the carpet. The stains should be covered by a towel for blotting them. This is repeated using dry towels until all the liquid has been absorbed. In the case of large spills, a wet-dry vacuum can be used to remove most of the spills before blotting.

Rug care is determined by the size, construction and material of the rug. Large rugs should be vacuumed to remove dirt, and if a rug is reversible, both sides should be vacuumed. If pets are kept at home, a vacuum may leave pet hairs behind. A stiff brush should be used to remove these hairs, brushing in the direction of the nap of the rug.

5.2   Curtains and upholstery material

The care labels fixed to the rear of the curtains should be checked before laundering. Selecting professional curtain cleaners is the best choice. Regular gentle vacuuming with an appropriate soft vacuum attachment can help the curtains to be dust-free [301,302]. The removal of dust from the folds will prevent the grit from corroding the fabric. The edges of the curtains should also be cleaned regularly as they are frequently touched and may accumulate dust from the window sills.

The curtains should not be touched by hand as dirty hands can spoil the fabric. Rather a draw cord should be used. Any marks on the curtain can be spot cleaned with a damp cloth. The curtain should be dried immediately after spot cleaning to avoid any water marks. The curtains used in the kitchen absorb cooking odors and splattered grease over time; hence, they should be cleaned with special care. The curtains made of cotton and linen can be machine washed. However, silk curtains should always be hand washed. Prior to washing any curtain, a small portion should be tested to see its behaviour [273]. Some delicate or silk curtains may bleed colour in the first wash. The use of cool water and a mild detergent is preferred for the curtains whether they are machine washed or hand washed to avoid shrinkage.

After the curtains are washed and dried, they need to be pressed to refresh the fabric and make the pleats look sharper. When ironing, they should be turned inside out and ironed on a low setting. The curtains in the living room or bedroom with embellishments should be washed and pressed with more care to avoid pucker and damage to the embellishment. The curtains should be rotated regularly, to wear out evenly. Curtains and blinds should be used to protect the upholstery from direct sunlight through windows. The loose threads should never be pulled out, but rather cut to prevent any damage.

The upholstery should be vacuumed regularly with a soft upholstery attachment. The upholstery fabric can be slightly discoloured by the action of dirt combined with body moisture [303]. Regular vacuuming will remove the grit that can wear the fabric. While washing the upholstery, the care instructions should be checked that they are attached to the sofa. For better results, a professional dry cleaning service should be used at least once a year to keep the fabric fresh.

The upholstery fabric should never be placed in direct sunlight as it can fade and damage the fabric.

5.3   Bed linen

The bed linens also need proper care for maintaining their aesthetics and other useful properties. While following any care procedure for bed linen, the care instructions should be followed. The bed linen should always be cleaned by hand or machine washed, rather than dry cleaned. The linen items should be washed in lukewarm or cold water. While machine washing the linens, a gentle washing cycle and mild soap should be used. The following instructions should be followed for machine washing:

- Linens should be pre-washed before the first use. While washing, separate the linens from the other items in the wash and separate the light and dark colours.
- Linens should be machine washed in warm water on a gentle cycle with a cold-water rinse. Overloading of the washing machine should be avoided as this can cause fibres to break down from excessive abrasion and agitation. Care should be taken to pre-treat any stains prior to washing.
- A neutral or mild detergent without bleaching agents such as chlorine or peroxide is recommended. Detergent should not be poured directly on the linens; rather, it should be added to the water as the washtub fills, or added as dilute solution. Unless linens are very

soiled, only half the recommended amount of detergent needs to be used. Linens should not be chlorine bleached as it can weaken fibres and cause yellowing. For white linens, oxygen-based bleach should be used.

- Fine linens can be machine dried on low heat. Damp linens should be shaken well before adding to the dryer. Tumble drying should be done for no more than 5–7 minutes on low setting. The linens should be tumble dried at cool temperature or by air drying. They should be removed from the dryer, when still slightly damp as over drying can make them stiffer, and damage the lustre and lifespan. After the removal from the dryer, the linens should be laid flat or hung on a hanger to become completely dry. Over-drying is the most harmful process for fabrics as it weakens the fibres causing shrinkage and pilling. Line drying outdoors is also a good option when possible. Remove the linens from the dryer when they are still warm to avoid wrinkling.

- Linen items do not need ironing unless they are really wrinkled. If needed, they should be ironed while the fabric is still damp or ironed with steam at a medium-to-hot temperature. For best results, white linen should be ironed on both sides, whereas dark linen on the wrong side only. In case of damask/jacquard fabric, iron on the reverse side first, then on the front side to bring out the sheen. Embroidered items should be ironed on the reverse side atop a towel to preserve the three-dimensional effect of the embroidery. A press cloth should be used to protect delicate lace and cutwork.

- Before storing the washable linens, they should be washed and dried. Linens should be stored in a cool, dark and dry spot. When storing in a garment bag with other garments, the other garments should be of cotton, linen or muslin. Add appropriate tags as needed. A lavender sachet can be placed with linens to keep them smelling fresh. Linens should be rotated every 6 months and stored in bags that do not emit fumes or permit moisture, which may damage the material.

5.4   Other items

5.4.1   Blinds, shutters and awnings

While cleaning these items, it is essential to check the care labels before going for any cleaning. Regular dusting or the gentle wiping of blinds over the front of the fabric can help to avoid any harsh cleaning. Plasticised materials and sunscreen blind fabrics can be wiped down with warm soapy water or mild detergents. While wiping, vigorous rubbing of the fabric should be avoided as it can cause damage. More care should be taken while wiping flock-back blind fabrics. The use of any type of cleaner on blinds or leaving the blind wet while wiping should be avoided. Any marks from the front of the fabric can be removed by using a wrung-out cloth and gently treating at the spot. However, spot cleaning with excessive pressure can cause water marks on the fabric.

Shutters should be dusted regularly and can be wiped down with a damp soft cloth. Use of abrasive cleaners or any scourers should be avoided, which can damage the finish of the timber. Scrubbing the fabric or the use of harsh soaps, detergents, solvents, other liquid cleansers or any bleach should also be avoided. Dirty marks and mildew should be immediately removed by gentle brushing and cleaning well with cold water.

5.4.2   Blankets

Most of the blankets can be cleaned by hand or machine [304,305]. Prior to cleaning, excess dirt and foreign materials should be removed from a blanket. The blanket should be placed in a mesh wash bag so that the straps, buckles, etc. are protected from machine damage. Before washing a blanket, mend or replace bindings, treat spots and stains and ensure that the detergent is made for cold water. The following points should be considered while cleaning the blankets.

- **Machine wash:** Blankets should be machine washed in cool water on a gentle cycle using mild soap or a blanket-cleaning formulation. Blankets that are machine washable should fit within the washing machine or they will not get washed properly. Softener sheets or liquid softener may be introduced as well.
- **Water temperature:** Cold water should be used for blanket washing, which will keep the blanket in good condition.
- **Drying:** Blankets should preferably be line dried and not tumble dried. Blankets should not be put in a hot dryer. A medium-heated dryer or permanent press cycle must be used for the best results.
- **Electric blankets:** These blankets should be washed carefully. They should never be dry cleaned, as the dry cleaning solvents can damage the electrical wiring. Similarly, mothproofing is harmful to the electrical wiring also.

5.4.3   Leather goods

Leather goods need to be cared for similar to the human skin as they are the tanned product of an animal [139]. Appropriate care and maintenance will ensure leather stays in good condition and increases its durability. The care instructions should be followed and the genuineness of the leather has to be confirmed before any treatment. Care practices for leather, suede, nubuck or synthetic leather should not be used interchangeably.

- Prior to cleaning, excess dirt and foreign materials should be removed from leather goods. Pre-treatment with a leather shampoo may be necessary before any conditioner or sealer is applied, depending on the surface and level of soiling. While applying a conditioner or sealer, it is wise to use a clean cloth before transferring to the leather, to assist even coverage and absorption.
- Annual cleaning and reconditioning is recommended for leather jackets. If the jacket has a smooth surface, the entire surface should be wiped using a mild soap or detergent. A damp sponge can be employed to rinse the jacket and blot excess water. The jacket should be dried in a place with abundant air circulation. Reconditioning of leather items is done to prevent the leather from drying out and getting damaged by the external elements. Conditioner and waterproofing products are applied in a similar manner to when the jacket was new.
- While removing stains, a spot test should be carried out before using any cleaning agent on the leather. This ensures that the leather will not be damaged. This is done by applying a small amount of the cleaning agent on areas such as the collar of the jacket and letting it set for about 10 minutes. The applied spot is checked to ensure that there is no damage to the applied area. In order to remove ink stains from leather, a non-acetone nail polish remover may be used.
- A leather jacket should always be hung up on a wooden or padded hanger. Using a wire or a thin plastic hanger can cause indentations or damage to the leather.
- Creases can be removed from leather goods by hanging in the bathroom and running a hot shower. The steam will cause the creases to relax. While doing so, the jacket should not be placed directly under the water from the shower. Alternatively, creases can also be removed by ironing (at a low setting) using a layer of heavy paper between the iron and the jacket.

5.5   Care based on fibre type

Taking proper care of the clothing can keep away the frustration of repairing or even replacing garments, and hence, can save unnecessary expenses. Similarly, washing the clothes when it is needed, rather than each time the clothes are worn, can also save time and prevent wearing of the garment due to frequent cleaning. The following section describes the care to be taken to lengthen the lifespan of the clothing.

- Clothing subjected to excessive cleaning can wear out sooner. Depending on the type of garment and climatic conditions, some garments can be worn more than once before they are cleaned. Hence, a decision should be made after wearing a garment if it needs cleaning. One should check it for any stains, lint, odour or any other factor that would necessitate a cleaning. If the garment appears clean and there is no bad odour, it should be reused again before washing. Jackets can be brushed with a cloth-brush on the shoulder and other areas followed by hanging them on a padded hanger.
- 'A stitch in time does save nine' was well said by Benjamin Franklin. Minor repair to small tears or holes when they appear can prevent major repair work later, which may become irreparable subsequently.
- The hanger used to hold a garment should be of good quality to support the load of the cloth. Fragile wire hangers may lead the garment to sag out of shape and finally the shoulders of the garment may be damaged.
- The clothes should be stored with proper folding to avoid wrinkles. Storing clothes for a longer period of time needs special care to avoid the attack of microbes during the storage, which may damage the cloth. It's disheartening to pull out your favourite outfit and discover a moth hole. If prolonged storage is needed, for example the winter clothes in tropical countries, they can be stored in isolated places so that more room is available in the wardrobe for other clothes.
- Ironing them again not only wastes your time, but also adds to the wear and tear on your clothes.
- Sweaters should be cleaned, folded and stored in drawers or in storage boxes.
- While cleaning the clothes, the care instructions on the garment label should always be followed. This will save the clothes from damage. The use of hot water during washing or hot air during tumbling can cause shrinkage in rayon, wool, silk and their blends. The shrinkage can lead to improper fitting. Furthermore, there is also the threat of colour fading when hot water is used.
- Some delicate items should be washed by hand or using cold or warm water on the gentle cycle.
- The stains should be treated as soon as possible, before they are solidified. Use a commercial product for stain removal and use the appropriate steps. If there is any issue related to the colour-fastness, the clothes should be tested in an inside seam before using on the stained portion. If the garment is given to a dry cleaner, the location and cause of the stains should be mentioned to them.
- The clothes should be sorted before washing. Wash fabrics according to light or dark colours. You do not want dark colours bleeding onto your light-coloured clothes. Washing delicate fabrics separately

from sturdy fabrics protects the delicate materials, which might be damaged by rubbing against coarse fabrics. Wash items that produce lint, such as terry cloth towels, separately for obvious reasons.

- The use of a right type of detergent in an adequate amount is very essential. Wrong detergent and an excessive amount may damage the cloth. By using the proper amount, the cost of detergent as well as water can be saved. There should not be an excessive amount of suds in the water as the detergents left in clothing can irritate skin and damage the cloth.
- The drying instructions should be followed while tumble drying. Similarly, drying in sunlight or under shade should also be done as per the instructions.
- While ironing, setting the right temperature is essential for the type of fabric. Excessive heat can burn the fabrics instantly. Pressing very delicate fabrics should be done with care; using a press cloth adds another layer of protection.
- The clothes should be selected as per the occasion. For example, wearing a good shirt while cleaning a car can damage or stain it. Hence, changing the shirt or other good clothes before this sort of work can save the clothes from potential damage.

5.5.1   Cotton items

Cotton fibre is often used as blends with other fibres so it will not shrink and wrinkle easily. Hence, the care instructions should be checked while washing for any special exceptions [3,201,306]. However, cotton clothes can generally be easily machine washed and dried. Best results can be obtained by using warm water, regular detergent (with colour-safe bleach if desired), normal wash cycle and tumble drying on a normal setting.

- Cotton items should be separated by colour and weight. Lighter colours can be washed in warm water, whereas dark colours such as denim, corduroy and canvas can be washed in cool water.
- Cotton whites can be washed with bleach on a hot water setting. Bleaches should be avoided if a finish has been applied to the garment. An excessive amount of bleach should be avoided as it can damage the fibres.
- Excessive drying of cotton clothes should be avoided, as it can cause wrinkles as well as shrinkage. The clothes should be removed from the dryer when they are slightly damp or allowed to air dry.
- The stain-removal quality of cotton is outstanding. Applying water, seltzer or an ice cube to the affected area can help in easy stain removal. Pre-soaking in detergent before washing can remove even really dirty stains.

- While washing cotton towels, use of a lower amount of recommended detergent with an extra rinse cycle can help to retain the softness and fluffiness as residual detergent can affect the softness. Similarly, silicone-based softeners should be avoided as they will diminish the absorbency of the towels due to their hydrophobic nature.
- A hot iron should be used to remove the wrinkles. The clothes should be turned inside out before ironing. Keeping the iron at one place of a cloth for a longer duration should be avoided. The crispness of cotton fabrics can be revived by adding a spray starch during ironing. Hang cotton shirts and pants to prevent wrinkling.

5.5.2   Woollen items

The natural soil release ability of woollen clothing enables the longer use between two consecutive refurbishing cycles. In some instances, washing can be replaced with airing or spot cleaning. Woollen clothing is generally subjected to shorter washing cycles and lower washing temperatures than other fibres. Hence the energy consumption is much lower than other fibres [307]. Wool is the potential natural fibre to reduce the environmental effect from textile care and maintenance, which is the most energy-demanding phase, and may exceed the production, transpiration and disposal phases [308,309]. However, woollen clothing is slightly expensive and its properties can often change due to improper care and maintenance practises.

Generally, woollen items are washed at lower temperatures. In order to achieve the desired disinfection at a lower temperature, a longer washing cycle and detergents with higher bacterial killing efficiency are needed [310]. The professional laundry-systems chemical disinfection (such as detergents containing peracetic acid) is used [311]. Hence, from a laundry worker's perspective, the care for woollen items is not hard. The following list describes the specific care and storage instructions for woollen clothing.

1. Wool fibre is used in knitwear manufactured mainly by a fully fashioned and flat knitting machine with some circular knit (wool book). The consumer demand for machine-washable wool knitwear and the modern requirements for easy-care wool that needs tumble dryability are increasing [312]. Furthermore, the increased use of domestic washing machines and tumble dryers has necessitated the research efforts to produce shrink-resistant methods for wool. It is a well-known fact that wool fibre is highly vulnerable to felting shrinkage, which is caused by the directional frictional effect (DFE) [313–315]. Any attempt to mask the scales of wool can

reduce the DFE and hence, the felting shrinkage. Processes such as chlorination-Hercosett and treatment with chlorine containing shrink-resistant polymers can be adopted to prevent the shrinkage. However, the chlorination processe is vulnerable to the problem of adsorbable organic halogen (AOX) compounds, which results in the reaction of chlorine with wool. The other processes are discussed in detail in the paper by Holme [297].

2. Cleaning of wool in a machine should always be done separately or it should be hand washed [316]. The woollen garments need low temperature and short-duration washing. A mixed load in the washing machine can cause the leaching of dyes from other fabrics and cross staining of woollens. It will be difficult to clean the stained woollen item.

3. Another reason for not washing woollen items with other clothing is the increased potential of pilling [317,318]. The mechanical agitation during washing can lead to pilling, which can be severe if washed with synthetic fibres.

4. The use of an appropriate detergent or dry cleaning chemical is very essential for woollen items. There are commercial products available for woollen clothing, however, the nature of the chemical mixture should be analysed while selecting an appropriate chemical. Although some of the commercial laundering agents use the name wool, they may not be appropriate for woollen items. The detergents or chemicals should be mixed with hot water first and then the solution should be mixed with a larger amount of cold water. This will help in better mixing of the laundering agents with water. If the laundering agents are not meant for woollen items, use as minimal an amount as possible of the detergent to avoid felting shrinkage.

5. Daily care activities such as brushing, airing and pressing are important especially for woollen garments. The same woollen garment should not be used on a daily basis for a longer time. The woollen garments need some time (about a day) between two consecutive wearings to regain their shape.

6. The woollen items labelled with a care instruction 'hand wash only' should be washed in cool water using a mild detergent. The use of bleaches should be avoided as they damage the wool fibre. Rinsing twice in cool water is preferred to remove excess detergent, and extra water should be removed by a gentle squeeze. The woollen clothing should not be rubbed against itself, which will cause felting shrinkage. High temperature, harsh detergent and agitation can lead to shrinkage and colour fading. In several instances, the woollen items can be easily cleaned with just cold water. The twisting or wringing should be avoided in order to retain the shape. They also should be dried flat to retain the size and shape.

7. In order to clean the woollen items that are labelled machine washable, the delicate washing cycle should be selected with the use of mild detergent. Both the washing and rinsing should be done either with cool or warm water to avoid felting shrinkage. The correct machine wash includes filling the machine with cold water, adding soap (if any), agitating the water to mix the soap, adding the woollens by pressing them down into the water, agitating them by hand very gently and pushing them under the water. After this, they should be left for about 30 minutes followed by setting the machine to the part of the cycle where it drains, then refills for rinsing without any agitation. Agitation with the presence of detergent can cause felting shrinkage.

8. The woollen items should be spin dried at a lower rpm (400–1000 rpm) to remove the excess water [307]. This indicates a higher amount of moisture remaining in the cloth, which takes more time to dry. It was established that the spin dry speed can be increased up to 1400 rpm without substantial shrinkage [307]. Two cycles of rinsing and spin dry are appropriate for the majority of clothes. Furthermore, they should be dried flat after washing to retain the size and shape.

9. Woollen garments that are of high quality, expensive or not used frequently should not be cleaned regularly. They should be cleaned after wearing the garment once or twice or a few more times depending on the use. This will help to maintain the natural resiliency and springiness of the woollen clothing. The items used frequently should also be washed as little as possible. If possible they should be managed with spot cleaning and airing so that the durability is increased.

10. Woollen garments that are large in size and difficult to handle should be sent to the commercial cleaner. Before sending them, all the worn places should be mended to reduce further damage.

11. As the woollen items are prepared from wool fibres that vary widely in their qualities, it is essential to follow the care instructions recommended by the manufacturer. The same clothing styles made from different wool fibres should follow different care procedures.

12. Drying the woollen clothing under direct sunlight should be avoided to retain the shade. Neither should it be dried in the dryer as the combination of heat, friction and pressure can cause shrinkage.

13. Avoid hanging woollen clothing after washing as it easily loses its size and shape.

14. After washing and drying, the woollen items should be allowed about 24 hours before wearing again. This will help to recover the natural resiliency, to remove the wrinkles and to retain the original shape.

15. If stained, the stains should be treated with water immediately while still wet to prevent settling of the stain on the fabric. Otherwise, a clean white cotton cloth can be used to absorb and remove as much

of the stain as possible. Harsh rubbing to the stained area should be avoided. If the stain is dried, it will be hard to remove. In this scenario, or with the stains that are hard to remove such as grease or ink, professional cleaners should be consulted. An improper attempt to remove grease stains often proves to be ruinous for woollen items.

16. Heavily soiled woollen items should be soaked in cold water with a mild detergent for a few hours. The soil mark should be removed gently by hand or by a soft brush to retain the colour and avoid shrinkage. The use of a hard brush or harsh detergent should be avoided.

17. For removing alcohol stains, the stained portion should be soaked in cold water and laundry detergent should be used at stained portions. A gentle scrub by hand or a soft brush can help to remove the stain.

18. If the stain is created by mud, it should be wetted first and then sponged with soapy water. In order to deal with perspiration stains, the stained area should be sponged with white vinegar.

19. For the removal of any type of stain, the stained portion or the whole garment should be immersed in sufficient quantities of cold water. An attempt should be made to remove the stains before immersing the whole garment, which can cross stain the other areas. The stains should be removed before pressing. Otherwise, heat can cause stains to settle in the wool fabric.

20. If any chemical is being used for stain removal, it should be safe for woollen clothing. The chemical can be tested in a concealed area of the garment before applying onto the stain.

21. The wrinkles on woollen clothing will be automatically removed if the clothing is kept hanging in a dry state for a sufficient amount of time. When the woollen items need to be pressed or ironed, the iron should be set on 'Woollens'. The garment should be sprayed with a little water prior to ironing. Woollen items should never be ironed dry. The clothing should be pressed in the inner side with a smooth motion avoiding excess pressure or rubbing, which can create shiny spots. If shiny spots appear after ironing, spraying a small amount of distilled white vinegar followed by rinsing with cool water can remove the shiny spot.

22. While using a heavy iron for top pressing, a clean white cotton cloth should be placed on the portion to be ironed to avoid damage or create shiny spots.

23. While pressing napped woollen items, a thick terry cloth towel should be placed on the ironing board to prevent crushing. A trial test can be done in a hidden area before pressing. If the napped item is scorched slightly, rub gently with an emery board. For

more severe scorches, a diluted solution of hydrogen peroxide (H_2O_2) can be used.

24. When it is needed to press a high amount of woollens, a steam iron can be used. The pressing of curved areas such as lapels and collars can be done by using a tailor's ham. Similarly, for pressing seams open without making a visible seam edge, a seam roll can be used. For pressing the areas hard to reach, a point presser along with a press cloth can be used.

25. Woollen clothing should be stored after cleaning in a dry container with moth balls to repel moths and insects. The direct contact of moth balls with the clothing should be avoided. They should be packed in loosely woven cloth bags and placed near the storage place. The food or other biological stains should be definitely removed because they attract moths and insects. The woollen items should be stored in a dry area and clean storage space. White tissue paper should be used between the folds (if folded for storage) to prevent wrinkling. When removed from the storage place, the woollen clothing needs to be aired out to remove the odour of the mothballs.

5.5.3   Silk items

Silk is a natural protein fibre similar to human hair. The clothing items prepared from silk fibre are highly sensitive to the temperature, type of detergent and bleach [319–321]. Therefore, while washing silk garments, appropriate care instruction should be followed, which can ensure durability of these items, as described below:

- Although machine washing is mentioned in the care label, hand washing of silk is always the ideal choice.
- While using a washer, use the delicate cycle, mild detergent and the shortest spin cycle.
- In a top-loading machine, use a mesh bag for extra protection (this step is not necessary in a front-loading machine).
- Avoid the use of perfume or deodorant for silk items.
- Avoid the use of bleach, especially chlorine bleach, for silk items, as the bleach can damage the fabric.
- Silk should never be exposed to direct sunlight for any longer than need be because it may deter the properties. Wet silk may turn yellow in direct sunlight.
- While drying, excessive temperature and heat can be detrimental to the lustre of the silk fabric. In addition, the friction with the dryer drum might cause yarn breakage.
- When using a dryer, the air dry setting should be used without any heat.

- Avoid the spraying of water while ironing.
- Individual stains should not be treated with water.
- Always use the iron in the backside of the cloth using a cool iron or setting for silk. Excessive heat can dull, pucker or burn silk fabric.
- Hang silk garments to dry as they will retain the shape.
- Minor silk wrinkles generally disappear if the garment is hung overnight.
- Major wrinkles can be removed by cool ironing.
- Hanging the silk garment in the bathroom during a shower can help in the removal of the wrinkles by the humidity.
- If any stains need to be removed, a professional dry cleaner should be consulted. The sooner the stain is treated, the better.
- Almost all silk garments can be easily hand washed without being damaged.
- A mild non-alkaline soap or specially formulated shampoos, such as a baby shampoo, should be used. Detergents that contain enzymes and brighteners should be avoided.
- Silk items should be pre-soaked in lukewarm water for 3–5 minutes. Pre-soaking prevents the silk items from shrinkage. They should be gently moved from side to side during soaking. However, dark or printed items should not be soaked. They should be washed in cold water.
- If there is hard water, the hardness can be removed by adding a spoon of borax to the washing water.
- After soaking for a maximum of 5 minutes, the silk items should be thoroughly rinsed with cold water by adding a teaspoon of vinegar to remove the soap completely.
- Adding a few drops of hair conditioner to the final rinse water can provide an extra silky feel.
- As silk is very delicate, the silk items should be handled with great care while they are wet.
- A dry towel can be used to remove the excessive water after washing.
- Roll out the silk and straighten it gently at the corners.
- Dry flat, never wring dry or twist.

5.5.4   Nylon items

The following care should be taken for nylon clothing items or its blends [322,323]:

- While nylon items are machine washed, cold water and a gentle cycle (as mentioned in the care instruction) should be used.
- Nylon clothes should be washed separately from other types of fabric.

- Any commercial laundry detergent can be used for nylon clothes.
- Only non-chlorine bleach should be used for nylon clothes.
- If a dryer is used to dry the nylon items, the dryer should be set on its lowest setting. The items should be removed promptly when the cycle is finished.
- Some nylon items can develop pills due to extensive washing and tumble drying. Hand washing and drip drying can be used to avoid this problem.
- Delicate nylon items such as lingerie and hosiery should be washed with the use of warm water and gentle detergent for best results. After washing, hang or lay flat to dry these items.
- The problem of static on nylon can be avoided by adding a fabric softener either during the rinse cycle in the washer or with a dryer sheet in the dryer.
- A cool setting should be used while ironing nylon items.

5.6   Problems during laundering and dry cleaning with solutions

Various problems such as wrinkling, shrinkage, distortion, colour loss, non-removal of soil, staining, change in texture and other changes in appearance are concerns in laundering and dry cleaning [324,325]. The appropriate selection of laundering chemicals and washing cycles and the attention to care instructions can avoid these problems. Various problems such as wrinkling, shrinkage, distortion, colour loss, non-removal of soil, staining, change in texture and other changes in appearance are concerns in laundering [325]. The appropriate selection of laundering chemicals and washing cycles and the attention to care instructions can avoid these problems. Table 5.1 describes the problems and necessary solutions during the laundering operation.

Similarly, there are certain things both the customer and dry cleaner should take care of to avoid problems during dry cleaning as discussed below:

- The dry cleaning should be done in accordance with the care label instructions.
- While accepting an article for dry cleaning, the dry cleaner should inspect it and get the history of the stains or damages. If the garment is damaged, depending on the circumstances, the dry cleaner may be responsible for the damage.
- The customers should check all the pockets prior to dropping off the garment to ensure that there are no foreign objects that may spoil the cleaning process.

Table 5.1 Problems during the cleaning of textiles and possible solutions

Problems	Probable causes	Solutions
Greying of clothes	Insufficient amount of detergent.	Increase the amount of detergent and/or use a detergent booster or bleach.
	Wash-water temperature too low.	Increase the wash temperature to a permissible higher value.
	Incorrect sorting, transfer of colour.	Separate heavily soiled items from lightly soiled ones; separate dark and light colours.
Uneven cleaning	Insufficient use of detergent after treating with a pre-wash stain remover.	Soak clothes in a concentrated solution of a liquid laundry detergent. Re-wash with an increased amount of detergent.
Yellowing of clothes	Insufficient amount of detergent.	Increase the amount of detergent and/or use a detergent containing enzymes, detergent booster or bleach.
	Wash-water temperature too low.	Increase the wash temperature to a permissible higher value.
	Hand washing synthetics with light-duty detergent. Treating synthetic fabrics in a 'delicate cycle'.	Use hot water (60°C) and a permanent press cycle. Increase the amount of detergent and/or use a detergent booster or bleach.
Fabric discolouration	Use of an incorrect bleach type.	Hard to remove yellowing. Hence, use the right type of bleach.
Poor washing	Insufficient amount of detergent.	Use a sufficient amount of detergent.
	Wash-water temperature is too low.	Increase the wash temperature to a permissible higher value.
	Overloading of the washer.	Reduce the wash load. Sort clothes by colour, fabric and the amount of soiling. Use a proper water level as per the load.

(Continued)

Table 5.1 (Continued) Problems during the cleaning of textiles and possible solutions

Problems	Probable causes	Solutions
Residual detergent or streaks of	Undissolved detergent.	Add the right type of detergent to the water first before adding clothes and starting the washing cycle.
powder present on dark or bright colours	The use of hard water that combines with detergent to form a residue.	Increase the wash temperature to a permissible higher value. Do not overload the machine. Use soft water, liquid laundry detergent or a non-precipitating water softener with a powder detergent.
Stiff, harsh fabrics, increased fabric wear and abrasion	The use of hard water that combines with detergent to form a residue.	Use soft water, a liquid detergent or a non-precipitating water softener with a powder detergent.
Lint	Improper sorting; mixing clothes such as sweaters, bath towels and flannels with synthetics, corduroys, velours and other napped fabrics that cause lint.	Sort the load and wash the clothes that form lint in separate loads from synthetic or napped fabrics. Very heavy lint shedders such as blankets, chenille bedspreads or rugs should be washed separately.
	Overloading of the washer or dryer.	Reduce the wash load and use a proper water level as per load size.
	An insufficient amount of detergent.	Increase the amount of detergent in order to hold lint in the solution during the wash time.
	A clogged lint filter.	Clean lint filter.
	Overdrying in a dryer that creates a build-up of static electricity in synthetic fabrics.	Use fabric softener in the washer or dryer to reduce static electricity of synthetics. Avoid overdrying.
	Dryer lint screen is full.	Clean lint screen after each use and dry with a cleaned lint screen.
Pilling	Pilling is caused in synthetics as the fibres break off the surface, ball up and cling to the surface.	Pilling cannot be prevented completely. It is a natural characteristic of some synthetic and permanent press fabrics. Use a fabric softener in the washer or dryer to lubricate the fibres. Use a spray starch or fabric finish on collars and cuffs while ironing.

(Continued)

Table 5.1 (Continued) Problems during the cleaning of textiles and possible solutions

Problems	Probable causes	Solutions
Holes, tears or snags	Incorrect use of sodium hypochlorite bleach.	Follow the guidelines on the use of bleaches. Never pour liquid sodium hypochlorite bleach directly on clothes. Use the bleach dispenser in the washer or dilute with at least four parts of water before adding to the wash water.
	Unfastened zippers, hooks and belt buckles that readily snag synthetic knits.	Fasten zippers, buckles, hooks and eyes before adding to the washer. Turn synthetic knits inside out while washing.
	Rips, tears and broken threads in seams.	Mend any visible damage before washing, especially open seams that will fray and become difficult to mend.
	Overloading the washer.	Let wash load circulate freely. Use the proper water level for the amount of clothes being washed.
	UV degradation [326].	Check items like curtains before washing by gently pulling the fabric to determine their condition. Use a gentle cycle for curtains that can be washed.
Colour loss or colour fading	Unstable dyes used in the garments.	Follow the instruction on the care label. Wash new items separately the first few times to remove excess dye. Sort load before washing.
	Improper use of bleach and incorrect detergent type. Rubbing with water will cause these colours to bleed or fade.	
	Use of too hot water for coloured fabrics.	Use cooler water for coloured items.
	Improper use of bleach.	Test item for colour fastness before using bleach. Use oxygen bleach.
	Undiluted bleach applied directly to the fabric.	Do not pour undiluted bleach directly on the clothes. Follow the instructions for the correct use.

(Continued)

Table 5.1 (Continued) Problems during the cleaning of textiles and possible solutions

Problems	Probable causes	Solutions
Wrinkling of synthetic or permanent press fabrics	Incorrect washing or drying cycle.	Use the permanent press cycle on the washer and dryer. Use warm wash, slower spin speed in the washer, and cold rinse. Remove items from the dryer as soon the cycle completes; hang or fold items.
	Failure to remove items promptly from the dryer at the end of the cycle.	Remove items from the dryer as soon the cycle completes; hang or fold items.
	Over drying.	Reduce drying time and remove items when there is a trace of moisture in them; hang or fold items.
	Overloading of the washer and/or dryer.	Do not overload the washer or dryer and use fabric softener.
Shrinkage	Over-drying.	Avoid over drying and remove the clothes when there is a trace of moisture in them. Stretch back into shape and lay flat to finish drying.
	Residual shrinkage.	Many knits and woven fabrics shrink when first laundered. Check the quality of the item and read care instructions.
	Agitation of woollen items leading to felting shrinkage.	Reduce agitation. Use a gentle cycle or soak method for washing and rinsing.

- The article should be fully examined by the customer as well as the dry cleaner for any damages, cloth blemishes, fading, tears, etc. noted on the docket.
- Customers should keep their dockets, and the dry cleaners should maintain good records in case a problem occurs in the future.

chapter six

Stains

Stains are local deposits of soiling or discolouration that exhibit some degree of resistance to removal by laundering or dry cleaning, thus creating critical issues in garment care [280,327]. The presence of stains on a garment makes it dull, stiff and vulnerable to attack by insects. Any attempt to remove stains may cause colour loss or abrasion. Stain removal is affected by the age, extent and type of stain and the type of fabric. The fibre content, fabric construction and the dye and finish characteristics should be considered before stain removal, as the same stain may respond differently in different fabrics. Failure of the cleaning method to remove soiling and stains may lead to product failure.

6.1   Types of stains

Stains can be classified according to their characteristics as water- or solvent-soluble and insoluble. They can also be classified according to the method of removal as protein stains (milk, blood, albumen, pudding, baby food, mud, cream, egg, gelatine, vomit and ice cream), tannin stains (beer, alcoholic beverages, coffee, cologne, fruit juice, soft drinks, tea, tomato juice and berries), oil-based stains (hair oil, automotive oil, grease, salad dressing, butter, lard, suntan lotion and face creams), dye stains (cherry, mustard and colour bleeding in the wash) and combination stains (candle wax, ballpoint ink, lipstick, shoe polish, tar, eye makeup, barbecue sauce, gravy, hair spray and tomato sauce) [328–333].

Identification of the type of stain is important to prevent its removal from damaging the fabric. As ageing and heat can set stains permanently, they should be removed as soon as possible [2]. The selection of an unsuitable method may also set the stain permanently. The removal of some stains requires special techniques and solvents and should be done by laundry professionals.

6.2   Removal of stains

The most important thing to be considered while dealing with any kind of stains is to prevent them from setting [2]. Once set, the staining material forms a chemical bond with the fabric, which is permanent and hard to remove. Removing the set stain can lead to the discolouration of the fabric and the rejection of the discoloured fabric itself. Sometimes excessive rubbing of a set stain by scrubbing can lead to the stained fibers being worn off, leaving the unstained ones visible. To prevent the rejection of a garment with a permanent stain, these general guidelines should be followed:

- Treat any stain immediately with water or with the appropriate solvent if it is available.
- Avoid direct heat, as heat will cause most types of stains to be set in the fabric. Stained clothing should not be placed near radiant heat sources; always room temperature or lukewarm solvents should be used.
- The solvents or water should be gently applied, dabbing them onto the stain and letting them soak in rather than scrubbing forcefully.

Polyester fibre and durable press (DP) finishes retain oily soiling and create cleaning problems [36]. The staining characteristics of resin-treated fabrics have been evaluated by Reeves et al. [334]. It was reported that the removal of soiling from a garment was affected by the fibre content, the type of soil and the process of producing cross-links of resin and catalyst. Polyester/cotton (P/C) fabrics with and without resin treatment soiled more readily with an oily soil and retained more of the soiling after repeated laundering than did similar cotton fabrics. However, with non-oily soil, the P/C fabrics soiled less than cotton fabrics and retained less soil after laundering.

Special care should be taken with regard to temperature during the removal of albuminous stains as higher temperatures may accelerate the coagulation of albumen and fix the stains [335,336]. Some stains such as glues and paints that contain epoxy resin as a base will damage the fabric during removal. Oily stains should be sponged with a dry solvent and non-greasy stains should be removed with water. Garments packed in polyethylene bags may stain when subjected to excessive heat, e.g., packing the heated garment just after ironing, or carrying garments in a car under high-temperature conditions. Ink stains require skill and specialised techniques for complete removal.

6.3   Impact of stain removal on clothing properties

The stain removal with chemicals and heat can alter the clothing properties [337]. The major effect can be colour fading on the portions treated with the chemicals. In some instances excessive rubbing can lead to abrasion or pilling or some other mechanical damage as discussed in Section 2.7.

chapter seven

Storage of clothing

7.1   Apparel textiles

The effect of storage on the performance of garments is of special interest to manufacturers, retailers and consumers. The problems associated with the storage of garments are insect, rot and mildew damage or other conditions that may create problems during subsequent use. When textiles are stored in damp or in highly humid conditions, they are vulnerable to insect damage. This will be aggravated if the storage spaces are dark and stagnant with warm and humid areas. Before storing a garment, the following points should be noted:

1. The garment should be cleaned and all stains removed.
2. All the fastenings should be closed.
3. Belts should be removed from their loops.
4. The garment should be hung from a coat hanger.
5. A moth preventive should be sprayed if the garment is vulnerable to insect attack.
6. The storage area should be cool and dry.

Many insect species may damage soiled textiles and those made of wool or other animal hair fibres such as mohair, angora and cashmere. Crickets and silverfish may cause irreparable damage to cellulosic garments and a distinct odour may arise, particularly when starch, glue or other attractive materials are present. If synthetic fabrics are stored in a dirty, spotted or stained condition, they may be damaged by insects. A reliable moth preventive such as naphthalene or paradichlorobenzene in the form of balls, flakes or powder should be applied before storing. Storage areas should be cool, dry and away from sunlight to prevent the hatching of insects.

Some expensive garments such as coats and furs should be stored in appropriate storage vaults. Furs must be stored at the correct degree of humidity, as high humidity can cause damage by mould or mildew, and low humidity draws moisture and natural oils from pelts and fur hairs, thus shortening the life of the fur. Knitted items and sweaters should be stored flat with tissue paper stuffing so that fold marks will not be obvious. If pile garments are folded into boxes, the pile may be distorted. During recent decades, an increasing number of storage firms have installed burglar-proof and fire-proof storage vaults. Some professional cleaners also provide wardrobe-storage hampers or box-storage and will store garments in a vault with controlled temperature and humidity.

Wool can easily lose its shape and become less defined if washed without care, and worst of all, it can shrink, becoming quite a few sizes smaller. Indeed, it is not unknown to pull out a child's sweater after putting in an adult's one! If you're not seeking such a transformation of your woollen garments, it makes sense to follow these instructions.

7.2   Storage of personal protection equipment (PPE)

Some of the contaminated protective clothing after use may pose a risk during the laundering process. In addition, selection of appropriate cleaning protocol is very essential, which can affect the performance in the subsequent use. Hence, the cleaning should be performed by a trained professional or by specialised laundries. Sufficient information on care and maintenance should be obtained from the manufacturer of the PPE.

The PPE should be properly cleaned before storage, which may extend the life of the item. The PPE should be appropriately cleaned and dried. When the PPE is not in use, it should be stored in accordance with the manufacturer's recommendations. The vaults should be checked and repaired as needed. The PPE should be placed in an appropriate, clean container and stored in a convenient, uncontaminated environment. Covering it with some plastic wrapping can prevent dust accumulating in the garment. Atmospheric conditions such as excessive heat, moisture, direct sunlight, dust, chemicals, corrosive atmospheres or the presence of organic vapours may considerably reduce the life span of some PPE. The PPE will start degrading by the exposure to ultraviolet (UV) light or photo-oxidation during their storage if the closet doors are left open and the garments are stored for a prolonged period.

Several dyes and specialty finishes in the PPE are sensitive to UV light including the material from which they are manufactured. The most vulnerable part is the outer layer (in a multilayer ensemble), which deteriorates due to the degradation of the material, finishing and coating. This can lead to insufficient protection (e.g., from fire), or can reduce the water/oil repellency and wind proofing.

The used and/or contaminated PPE should be cleaned after use so that it is in a hygienic fashion in the next use. The users of the PPE should be trained on the appropriate use, care, and maintenance of the clothing to retain the protection properties. A record should be maintained on the care and maintenance of the PPE.

The protective gloves must be cleaned from the contaminants before storage. They should be stored away from high temperatures and direct sunlight in accordance with the manufacturer's recommendations. Some gloves may be affected by the presence of moisture and artificial lighting. Gloves for chemical protection should be rinsed in warm water to avoid the contaminants being dried on the gloves.

Protective helmets and harnesses should be cleaned with warm water and soap following the manufacturer's instructions before being stored. The storing of the helmets in hot places such as vehicle dashboards should be avoided.

Whenever possible, the PPEs used by a visitor should be disposable. If not practical, the non-disposable PPE may be used by the visitor, cleaned and stored as per the manufacturer's instructions. The re-usable respirators must be stored in a sealed container.

Insulating mats, barriers and covers need to be washed at intervals not exceeding 6 months and tested as per Australian/New Zealand standard (AS/NZS) 2978 at intervals of 6 months or less. They should be stored and handled carefully, away from chemicals, sharp objects, tools and other equipment to avoid damage. The electrical-insulating gloves should be washed and tested at intervals of 6 months or less and stored unfolded in clean containers in a cool and dry place away from direct sunlight.

chapter eight

The environmental impact and health hazards of cleaning

The use of non-aqueous textile cleaning has been in practise for several years. The properties of the textile and the nature of the soiling determine the medium for cleaning. The medium of cleaning in this case can be non-polar liquids such as petroleum, carbon dioxide or other similar liquids, which meet the textile cleaning performance requirements. The use of water and perchloroethylene (PCE, or perc) is not included in the non-aqueous cleaning.

8.1   Environmental impacts

The current global trend is to reduce the environmental load by the consumption of lesser materials, adopting alternative techniques, recycling and the reuse of material [338]. There are several things that can reduce the burden on the environment such as using the optimum washing load, lowering the washing water temperature, reducing the frequency of washing or devising a method of spot cleaning, using eco-friendly chemicals and eco-programmes, avoiding tumble drying and adopting alternative freshening methods such as airing [101,339,340].

As mentioned earlier, perc has been classified as carcinogenic to humans. Therefore, perc should be used as a hazardous substance during its storage and handling. The wastes derived from dry cleaners using perc should also be treated carefully and should not be directly discharged to any water course. The direct release of the perc fumes to the air can lead to the formation of smog by the reaction with some other volatile organic substances.

A home clothes dryer can generate about 2 kg of CO_2 per load of laundry dried. Hence, the dryers should be used only when it is necessary. Manufacturers of clothes dryers should measure the energy efficiency and improve it. Dryers should be labelled with energy consumption labels according to the amount of energy consumed per kilogram of clothes (kWh/kg). Automatic dryers can sense when the clothes are dried and switch off to save electricity and over-drying.

8.1.1   Chemicals with potential hazards

Some of the chemicals used in the manufacturing of detergents, soaps or other washing aids or dry cleaning chemicals pose an environmental threat [340,341]. Table 8.1 summarises the list of chemicals used in soaps, detergents and dry cleaning chemicals and their potential environmental impact.

Table 8.1 List of cleaning chemicals and their environmental/health hazards

Types of cleaning	Chemicals used	Environmental/health hazards
Laundering	Soaps, sodium carbonate or bicarbonate solution (neutral or weakly alkaline condition). Detergents: mixtures of anionic and non-ionic surfactants (e.g., alkali sulphates, fatty alcohols, alkylphenolethoxylates).	Many of these chemical formulations are a potential hazard when dispersed widely into the environment contaminating our soil, water and air.
Dry cleaning	Perc (most widely used).	A major source of toxic air pollution produces hazardous waste in many countries. Perc can contaminate soil and groundwater where they are disposed of. For these reasons, it is important to reduce and eventually eliminate this chemical from routine use [120]. High-level exposure to perc can affect the central nervous system, kidneys and liver, and cause mood and behavioral changes, impairment of coordination, dizziness, headache and fatigue. Chronic exposure to lower levels of the chemical can lead to cognitive and motor functioning impairment, headaches, vision impairment and in more isolated cases, cardiac arrhythmia, liver damage and kidney effects. Perc has also been demonstrated to have reproductive or developmental effects, and may cause several types of cancer.

The European Union has banned the detergents with non-ionic surfactants nonylphenolethoxylates (NPEs) and nonylphenols since 2005. However, NPEs are not being banned so far in the United States. Canadian facilities have planned to reduce the NPEs in the waste water [342].

Two recent studies at Georgetown University (in Washington, D.C.) have investigated that perc is retained in dry cleaned clothes and the amount of perc increases with repeated dry cleaning cycles. The second study shows perc, classified as carcinogenic to humans by the United States Environmental Protection Agency (EPA), is retained in dry cleaned clothes and that levels increase with repeated cleanings.

The alternative to perc is the use of liquid carbon dioxide (CO_2) for dry cleaning, which has been used for cleaning in other processes for about the last 3 decades or so [343–345]. For dry cleaning purposes, liquid CO_2 is formulated with additives and delivered to dry cleaners in pressurized canisters [346]. During the dry cleaning process, clothes are immersed in liquid CO_2 contained in an enclosed cylindrical basket (inside a pressure vessel that has pressures of 700–1000 psig). The load is then agitated inside the basket by high-velocity fluid jets or by mechanical action to remove soils. Once the cleaning cycle is complete, the pressure is released from the vessel, liquid CO_2 is vaporized and dry garments are removed.

It has been claimed that better cleaning is possible with liquid CO_2 as it has a low viscosity, which will help in the removal of smaller particles from the surface with less re-deposition [347]. In addition, liquid CO_2 is a non-polar solvent that is most effective in removing non-polar soils such as oily stains and greases. While experimental laboratory studies on soil removal from garments in liquid CO_2 appear promising, they have not yet been demonstrated under commercial conditions. However, it is inferior to perc in removing some forms of stains such as grease.

It was found that during liquid CO_2 dry cleaning, the level of the mechanical action has no influence on the removal of relatively small particulate and non-particulate soils [348]. Therefore, increasing the mechanical action cannot improve the washing results for relatively small particles (like carbon black and clay) that are received using perc. The use of suitable surfactants that reduce adhesion forces can help in the removal of relatively small particulate soils.

8.1.2   Recent innovations in laundering

Consumers always demand products that are better, cheaper, faster delivered and easy to care for. Furthermore there is a global trend to lower the environmental load. This has led to the research and development of new textile products such as durable press, wrinkle-free finishes, the creation of ultra-hydrophobic surface treatments and self-cleaning finishes.

8.1.3   Green cleaning

Green cleaning refers to the use of cleaning products and methods that are eco-friendly to preserve human health and environmental quality. Green cleaning products and techniques avoid the use of toxic chemicals, some of which emit volatile organic compounds negatively affecting the respiratory, dermatological and other health conditions in addition to the environmental pollution [349]. Consumers are also becoming aware and emphasising the use of eco-friendly cleaning products.

Conventional detergents can contain toxic ingredients that are harmful to the ecosystem, human beings and the environment [134,350]. For example, phosphates in conventional laundry soaps can cause algal blooms, which negatively affect ecosystems and marine life [89]. Hence, consumers should always select eco-friendly detergents. There are many commercial detergents available with labels indicating various eco-labels such as the product is readily biodegradable, phosphate-free and made from plant- and vegetable-based ingredients, which can help consumers to select an eco-friendly chemical. These detergents or soaps are gentler on skin and to the cloth.

Similarly, soaps are available that are made from certain tree seeds or oils that are easily biodegradable. A cup of white vinegar can be added to the washer during the rinse cycle, which acts as a softener. Vinegar naturally balances the pH of soap, leaving the clothes soft and free of chemical residue.

Various agencies are also focusing on the use of eco-friendly methods and products for laundering. For example, the EPA's 'Design for the Environment Program' labels products that satisfy the EPA's criteria for chemicals. These products can contain the 'Design for the Environment (DfE)' label. Generally, products labelled as 'low' or 'zero' volatile organic compounds (VOCs) are safer for human health as well as for the environment. The following describes the steps that domestic consumers can follow to make the cleaning as green cleaning:

Selection of concentrated detergents: As concentrated detergents use reduced packaging, there is a smaller carbon footprint and more products can be shipped at a cheaper price. The composition of the concentrated detergent should be appropriate for the clothing. These detergents can produce excessive foam in front-loading machines, which can damage them.

Wash by hand: Although hand washing is time consuming, there are certain advantages to it. This method is cheap and efficient. Hand washing gives you an idea of the laundry load on a weekly/daily basis and saves energy. Many clothes are being prevented from damage due to the rigorous process of machine washing.

Maximize energy efficiency: The washing machines that are too old consume excessive electricity as well as water. For example, a top-loading washing machine from the last century uses twice as much water per load than a newer machine. Hence, they can be replaced with the newer machine designs. However, if a replacement cannot be done, the following steps can improve the efficiency of these machines: (1) The use of cold water can save electricity consumption. About 90% of the energy used for washing clothes is used to heat the water. (2) With the availability of different varieties of detergents for cold-water washing, a huge amount of money can be saved with the use of cold water. (3) The machine should always be run with only full loads of laundry, which ensures maximum efficiency, or a smaller load option can be selected (if available in the machine) to save water and energy.

Hang it out to dry: The drying process consumes more electricity than the laundering process. Selecting outdoor drying or line drying can reduce the carbon footprint. The durability of the clothes increases with this method as there is less wear and tear than when a dryer is used.

Maximize dryer efficiency: When the dryer is used regularly, cleaning the lint filter after each drying cycle will increase the dryer efficiency and shorten the drying time. A moisture sensor in the dryer can reduce the amount of drying time or turn off the machine when the clothes are dry. Excessive drying, too high of a load and use of a damp cloth should be avoided to increase the dryer efficiency.

Avoid ironing if not needed: Ironing also consumes substantial energy and can deteriorate the fabric. In many instances it can be omitted by simply hanging clothes immediately after the wash cycle is complete. The water in the cloth will work with gravity to pull most wrinkles out. Then fold dry the clothes where you want the creases to be, and place them under other clothes in your dresser, which will further help to press them.

Use commercial services if needed: Commercial establishments are generally more efficient than home washing. Some special garments needing commercial laundering and dry cleaning should not be cleaned at home for maximum efficiency, proper care and maintenance.

Avoid chlorine bleach: Chlorine bleach may cause skin irritation and redness. Its fumes can irritate eyes, noses and airways, and if swallowed it can be fatal. Chlorine also poses a hazard as it can react with other washing aids to form toxic gases. If mixed with chemicals containing ammonia, it can produce lung-damaging chloramine gases. Chlorine mixed with acids, such as those in some toilet bowl cleaners, can form toxic chlorine gas, which can damage our airways. When released into the waterways, chlorine bleach can create organochlorines that can contaminate drinking water. Organochlorines are carcinogens that can damage reproductive, neurological and immune-system toxins. It has been known as one of the most enduring compounds to cause developmental disorders. Once introduced into the environment, it can take years, or even decades, to break down to less-damaging forms.

8.1.3.1   Ozone laundering

Ozone laundering systems have recently been shown to be successful in commercial installations because of the reduction in the use of energy, water and chemicals. Ozone laundering is eco-friendly and it has been proven that fabrics last longer, thereby reducing replacement costs. Ozone enhances the effectiveness of the chemicals by supplying oxygen to the laundry water, thus reducing the need for high-temperature washing with lower amounts of laundry chemicals [240,351]. Ambient to warm water temperature is needed for ozone laundering. High temperature dissipates ozone prematurely, negating its power, whereas, low temperature has a higher saturation level of ozone, providing better cleaning efficiency.

As ozone laundering systems normally require fewer rinse cycles, water usage is reduced by an estimated 30–45%. These systems recover most of the water used, so the reductions in water usage may be as high as 70–75%. Ozone oxidises the soiling in linen, making it easier to remove from the washed water. It can also reduce the need for harsh, high pH traditional chemicals for the same cleaning effectiveness. Ozone reduces the quantity of the chemical usage. The cost of chemicals is typically reduced by a minimum of 10%, but in certain cases the cost can be reduced by 50% [282]. Heavily soiled loads consisting of oily rags, food and beverages can be washed effectively with ozone in warm temperature water to get the quality-required cleaning. Typical reductions of energy are in the range of 80–90% for most laundries. The resultant savings claimed by laundries range from 5–30%.

New chemical formulations specially developed for low temperature ozone laundering are commercially available. Compared to the standard formulations for washing, these new chemicals greatly enhance the cleaning efficiency. Ozone in a water solution performs some of the functions of chlorine bleach. It assists in water softening by helping to remove cations such as calcium and magnesium from the water.

Ozone laundering improves the life and quality of textiles because it enables a shorter cycle time and a lower temperature. The wash cycles can be reduced by 10–40% per load as the wash and rinse cycles are reduced. A reduction in the amount of chemicals used also helps to improve the fabric life. Generally, the cost of linen replacement is much higher than water, chemical and energy costs. Ozone laundering can provide significant savings in this area.

In addition, the effluent will contain lower levels of biochemical- and chemical-oxygen demand (BOD and COD) because ozone oxidises bacteria, other micro-organisms and some dissolved organic compounds. The reduced washing and rinsing time means the laundry equipment is used more efficiently and the total staff hours per load are reduced. Furthermore, ozone is not dangerous to humans in the concentrations typical for ozone laundries compared to other chemicals.

8.1.3.2 Ultrasonic cleaning

Water is used for many laundering activities due to its high solvency for many substances in addition to the demonstrated occupational and environmental safety. However, the cleaning of hydrophobic contaminants is difficult to remove by water from hydrophobic surfaces such as polyester. The use of detergents and the mechanical action during washing can damage the fabric. The use of ultrasound, another type of mechanical action, for the cleaning of hard surfaces has been reported to be effective for textiles as well [194,352–355]. The application of ultrasonic energy can prevent the fabric damage that occurs in the conventional mechanical washers. The cleaning by ultrasound depends on the rapid formation and violent collapse of bubbles or cavities in cleaning liquids [356]. Ultrasound can be applied to textile laundering and dry cleaning operations.

The use of ultrasound cleaning is gentler to the fabric compared to the mechanical agitation of the laundering machine. The aggressive flowing and rotational agitations of wash bath and the deformations of the fabric and friction between the fabric surfaces do not take place in ultrasonic energy. Furthermore, ultrasound can remove the soils deposited in the fabric even at low liquor ratio and in a shorter cycle. This can lead to both water and energy saving. In addition, high-frequency ultrasound can remove stains on fabrics via a generation of active species in liquid. However, ultrasound is generally not suitable for regular cleaning of soft and flexible materials such as textiles.

The mechanism of cleaning by ultrasound is significantly different from conventional home laundering. In home laundering by machine or even hand washing, rubbing to the fibre surface occurs by aggressive stirring, agitation and plunging actions in the bath accompanied by chemicals. The transfer of the stains from the fibre surface into the bath is achieved with the help of detergents. However, ultrasonic laundering is a more gentle process, where the fabrics lie almost stationary in the ultrasonic bath. The acoustic cavitation produces a micro level of high-speed liquid movement accompanied by a vibration at the liquid/fabric interface through which the stains are removed. In theory, the deformation and damage to the fibres and yarns in a fabric induced by ultrasonic agitation may be much less than that of mechanical agitation.

Conventional laundering of silk items can lead to crease and deformation. The fabric rubbing during the laundering process causes fibre fractures in the form of fibrillation and degradation [320]. The appearance and wear endurance of silk garments are affected by this, which lead to considerable inconvenience in everyday use. As silk fabric can change in size and lose the hand property, the durable press finishing that is effective in improving the crease resistance to laundering, cannot be applied to silk. Dry cleaning of silk fabric is relatively ineffective in removing water-soluble stains and it is expensive. Ultrasonic laundering was found to perform much better in removing common stains from silk fabrics while the fabric appearance and dimensions were maintained compared to machine laundering [321]. In addition, there was less fibre damage by ultrasonic laundering to the fabric. This research showed that the introduction of ultrasonic laundering could produce a significant benefit for silk fabrics. However, ultrasonic agitation caused a slightly higher colour fading than the mechanical agitation. The K/S value (the ratio of absorption coefficient (K) over the scattering coefficient (S)) of the fabric laundered by ultrasonic agitation after 15 laundering cycles was only 2.7% lower than the fabric laundered by a washing machine, and the difference was not significant.

Blood stains from polyester/cotton-based medical surgery gowns [354] and soils from polyester and cotton fabrics can be efficiently removed by ultrasonic laundering in shorter time and at a lower bath ratio [357]. Home laundering of wool fabrics generally lead to felting shrinkage. However, ultrasonic laundering of woollen clothing can reduce fabric felting whilst achieving good stain removal compared with hand washing [194,352]. The repeated ultrasound washing cycles did not affect the colour and the tensile strength of the fabric [358]. Millions of bubbles or cavities are created by the ultrasonic energy with very high frequency into the liquid, which constantly strike at the target material surface, and as a result, remove the dirt off textile materials. The most important parameter in the mechanism of ultrasound cleaning is the power of ultrasonic cavitation in liquids. Ultrasonic use works better for the stain removal from a stained fabric. A higher velocity of the laundering solution at the fibre surface is achieved by ultrasonic irradiation. This increased velocity improves the mechanical removal of the stain from the surface of the fibre.

Ultrasonic cleaning has many advantages over the conventional laundering process such as (1) superior cleaning properties, (2) reduction in cycle time, (3) reduced energy consumption, (4) less chemical and (5) less mechanical damage to the clothing due to less fibre migrations. In spite of the above-mentioned advantages, there are several disadvantages associated with the ultrasonic cleaning of textiles, which are mentioned below:

1. It is difficult to achieve a homogeneous distribution of the acoustic field in the whole washing load, which may lead to irregular washing. There will be some areas of low acoustic energy that can result in improper cleaning. However, this problem can be overcome by moving the washing load continuously so that all the clothes pass through the areas of high ultrasonic intensity.
2. For the effectiveness of cleaning, the wash load has to be exposed to the ultrasonic field in such a way that there is the continuous production of 'strong transient cavitation' on the fabric surface. The collective cavitation of a large number of bubbles in the gassy liquid is

needed for good cleaning. However, the softness of the fabric facilitates cavitation to produce a small erosion effect while its reticulate structure favours the formation of layers of big bubbles that obstruct wave penetration [359].

3. The commercial development of ultrasonic washing machines for domestic laundering is difficult due to the difficulties in the designing of a suitable washing machine.

8.1.3.3   Eco-friendly chemicals for laundering

There are several companies producing environmentally friendly formulations for cleaning, which are easy to use, effective and productive in cleaning performance, and save resources and water. Many of them use detergents of non-petrochemical origin or substances that have been produced through genetic engineering [360,361]. The use of a washing substance closest to nature (such as soap) can guarantee a simple return into the natural cycle. Natural products such as baking soda, lemon juice, salts, pure soaps and palm oil can be used to make the products eco-friendly. Various companies are using these products to prepare formulations that include an ecological bleaching agent and stain remover, ecological washing powder, ecological laundry detergent, wool and delicate laundry detergent, soap bars and disinfectants. Consumers need to be aware of these products to make the care and maintenance of clothes eco-friendly.

8.1.3.4   Eco-friendly chemicals for dry cleaning

There are some chemicals listed as eco-friendly and are being used in the dry cleaning of clothing items as discussed below.

1. **Hydrocarbon:** Hydrocarbon-based solvents such as Chevron Phillips' EcoSoly or Exxon-Mobil's DF-2000 can be used in standard dry cleaning. These petroleum-based solvents are less aggressive than perc, hence, they need higher temperatures and longer cycles during dry cleaning. Hence, some heat-sensitive materials such as wool or acrylic may be adversely affected by the dry cleaning. Longer cycle time indicates more mechanical action on the fabric. Unlike the perc, these solvents are safe to beads, plastic buttons and sequins. However, some polyvinyl chloride (PVC) materials may deteriorate if progressively treated with hydrocarbons, and some adhesives may also dissolve. The other disadvantage of hydrocarbon is it takes longer to evaporate from bulky parts of clothing such as shoulder pads. It can leave small traces of oily residue that can cause allergies to some people. Although the hydrocarbon solvents are classified as combustible, they do not pose a high risk of fire or explosion when stored and handled properly. However, they can contain traces of volatile organic compounds that can form smog.

2. **Modified hydrocarbon blends (pure dry):** These are compounds prepared from hydrocarbons by modifying their properties.

3. **Glycol ethers:** Glycol ethers (dipropylene glycol tertiary-butyl ether) in several instances are more effective than perc and are eco-friendly. Various brand names for these products include Rynex, Solvair, Caled Impress and CaledGenX. One of the glycol ethers, 'Dipropylene glycol tertiary butyl ether (DPTB)' has a flash point well above the industrial specification. The ability of removing stains (water soluble) is equivalent or better than perc and other glycol ethers. The waste of DPTB can be easily separated by azeotropic distillation at a lower boiling point.

4. **Liquid silicone or siloxane:** Liquid silicone (decamethylcyclopentasiloxane, or D5) is a colourless, odourless and non-oily fluid that can be used for dry cleaning with a more gentle action and without any colour loss [362]. The popularity of siloxane is growing as it is considered a green solvent. As siloxanes hardly react with the textiles, the textile items retain their colour and quality. The use of liquid silicone needs licensing as it is in the list of the property of 'GreenEarth Cleaning (GEC)'. Although it is eco friendly, the cost is almost double of perc and GEC needs to be paid for the annual affiliation fee. It produces waste that is non-toxic and non-hazardous. It can degrade within a few days in the environment in the presence of silica and traces of water and CO_2. The research on female rats by Dow Corning established the fact that exposure to the solvent can increase the incidence of tumors. However, the male rats were not affected by the exposure. Further research carried on humans established that the threats observed in rats are not relevant to humans in this case due to the differences in the biological pathways.

5. **Liquid CO_2:** The use of liquid CO_2 is found to be superior to the commercial methods of dry cleaning using perc [363]. However, the cleaning efficiency of liquid CO_2 is fairly low compared to perc. The equipment used for liquid CO_2 is more expensive than perc equipment, which makes it difficult to be afforded by small or medium businesses. The environmental impact of CO_2 use is very low as the clothing does not emit volatile compounds. In addition, CO_2 cleaning can also be used for fire- and water-damage restoration as it is effective in the removal of toxic residues, soot and associated odours of fire. CO_2 is non-toxic, it does not persist in clothing and its greenhouse gas impact is lower than that of a majority of organic solvents. Some commercial dry cleaners use enzymes, which are equivalent to liquid CO_2 and more environmentally sustainable.

6. **Brominated solvents:** These are n-propyl bromide-based solvents with a higher Kauri-butanol (KB) value than perc (a KB value is an international terminology used to measure the solvency power of

a solvent based on hydrocarbon, which is specified in the ASTM standard: ASTM D1133). The higher KB value helps in the faster cleaning than perc. However, improper use can damage the buttons, sequins and beads. It is available in trade names such as Fabrisolv and DrySolv. In dry cleaning, the exposure to the solvent does not pose any health risk. Therefore, it is being approved by the EPA as a significant new alternative solvent compared to the traditional hazardous solvents. However, excessive exposure can lead to numbness of the nerves. Although the solvent is expensive, the cost is counterbalanced by its faster cleaning action, lower temperature and quick dry times. The overall dry cleaning cost per garment is found to be the same or lower compared to perc.

8.2 Health hazards

The exposure to laundering and dry cleaning chemicals poses various health hazards [118,140,364]. For example, the exposure to perc even at lower levels can lead to fatigue, headaches, dizziness, confusion, nausea and irritation to skin, eye and mucous membranes. The degree of these problems is related to the amount and concentration of perc and the duration of the exposure. Exposure to high levels of perc, even for short time, may cause serious symptoms, such as liver damage, respiratory failure or other fatal consequences.

8.2.1 During laundering or dry cleaning

Laboratory studies on animals have indicated that exposure to high levels of perc can impede the growth of a foetus and can cause birth defects and even death [126,365,366]. However, studies on people exposed to high levels of perc are either limited or inconclusive. It is yet to be determined whether exposure to perc can cause the adverse effects such as miscarriages or affect women's fertility or affect the child in the womb. However, it is an established fact that the dry cleaning chemicals can cause health disorders.

The people working in dry cleaning agencies can face serious health hazards as they spend a lot of time in an environment where the perc levels in the air are usually higher than outdoors [125,367,368]. Depending on the nature of the machineries and work practises, the level of exposure can vary from shop to shop. There are several ways in which perc is mixed with air as it evaporates rapidly. The various causes include:

1. Poorly maintained equipment.
2. Leakage in the equipment.
3. When the container of the perc is open to the air (such as while perc is added to the machine or transferred to other containers).

4. Perc waste materials after dry cleaning.
5. Garments not dried completely or processed improperly.
6. While transferring the clothes from a dry cleaning machine to a dryer in older machine designs (with separate washers and dryers).

The health hazards of perc exposure can be significantly reduced or even eliminated with improved technology, new dry cleaning equipment design and improved cleaning practices. For example, new machines, which clean and dry garments in a single unit, eliminate the transfer of wet garments from a washer to a dryer. Hence, they can significantly reduce the amount of exposure. Although the new equipment design and advanced technology can help to reduce the amount of exposure; improper work practises, inappropriate material handling and storage and improper maintenance of equipment can increase the exposure.

The potential of perc exposure leading to cancer has been extensively investigated [125,369–371]. The laboratory studies on rats and mice have established that perc can cause cancer in these animals when they swallow or inhale it. Several studies on people working in laundry and dry cleaning businesses have established the fact that perc exposure can pose elevated risks of certain types of cancer. The potential for an increased risk of cancer depends on factors such as the amount and concentration of perc exposure and the duration of the exposure. Furthermore, the individual's age, lifestyle, overall health condition and family traits also affect the risk of cancer.

8.2.2   Residual amount left in the clothing

It is the duty and responsibility of the professional cleaners to remove any traces of perc from dry cleaned clothes. A mild odour of perc is not a conclusive fact for the presence of any residual amount left in the clothes [372,373]. However, the dry cleaner can be approached to confirm that the perc has been removed completely or can be requested for a re-process of the garment if the solvent has not been completely removed.

The level of perc that the consumers of dry cleaning are exposed to by wearing the dry cleaned garments is not expected to cause any serious concern to an average person's health. However, within the premises of the dry cleaning services, the level of perc may be slightly higher than the outdoor air. Hence, frequent visits to the dry cleaner can cause minor effects such as skin irritation or nausea.

Several medical websites assign one of the main causes of dermatitis to the detergents and other chemicals applied to clothing during its cleaning. It has been shown that the fibres with irregular cross-sections are more likely to retain the detergent and other washing chemicals. If the rinsing process is not very effective there will be higher amounts of residual chemicals. For example, the bean-like cross-sectional shape of cotton fibres [374] may provide easy sites for the detergent residues to bind, and with the low uptake of moisture into the garment overall, could potentially cause retention issues. The fibres with a circular cross-section such as polyester/nylon may not retain much of the chemicals. Similarly, the staple yarns have higher chances of retaining these chemicals compared to filament yarns. The dyed and printed textiles are composed of a variety of chemicals that can form complexes with the detergents and other washing aids.

During high physical activity or in hot climatic conditions, the human body faces heat stress and therefore, perspires a lot to maintain thermal balance. This heat stress may be accentuated if chemicals from the dyes and finishes in the fabric leach out and become locally concentrated. In addition, there may be traces of detergents remaining in the fabric due to poor rinsing during laundering. All these factors can lead to skin irritation.

It has been observed that the use of proper washing conditions can help to remove any traces of chemicals from the wash load. Although commercial surfactants are normally used in permissible concentrations, it has been observed that the length of the alkyl chain in anionic surfactants is closely related to skin irritability. Compounds containing saturated alkyl chains of between 10 and 12 carbon atoms exhibit the worst effects [375]. Research on the allergenic properties of surfactants has shown that they pose no increased risk of allergy. The residual chemicals present on the cloth due to the poor rinsing can cause instantly occurring reactions of the skin, particularly in patients with pre-existing skin diseases. The reaction from these residual chemicals may cause stinging or itching [376].

8.2.3   Cross-contamination of diseases

There have been reports of infections among groups of people by the transmission of a virus due to the mixing of uniforms during laundering [377–379]. For example, methicillin-resistant *Staphylococcus aureus* (MRSA) infections among players of competitive sports and *S. aureus* infections transmitted by the wearers of protective clothing. Hence, appropriate cleaning of these uniforms after each use is highly recommended [380]. However, there is no specific evidence of such contamination reported by the users of cold-weather protective clothing. Still, there is always the risk of cross-infection by sharing the helmets and gloves. The threats can be higher in winter climates where low humidity upsets skin hydration, as abraded and chafed skin could be susceptible to *Staphylococcus* infection.

chapter nine

Future trends

The care and maintenance procedures have undergone several changes in the past decade or so, which affects the properties of apparel, protective and other textiles. Some of the changes include the formulation of greener soaps and detergents, new eco-friendly solvents for dry cleaning, the design and working of laundering and other similar equipment and the technical changes in the equipment to operate the new green chemicals. To cater to these changes, the fabric and garment manufacturers, in addition to their raw material suppliers, should make sustainable products as per the environmental sensibility [381].

Care labels always play an important role in the appropriate care and maintenance of many textile products. The durability, aesthetic values and dimensions of these items can be altered if the processes, process conditions and chemicals needed for care and maintenance are wrongly selected. Hence, the manufacturers should always include the right parameters in the care instructions and the consumers should follow them.

The main difficulties associated with care labels are: (1) some indicate procedures that are far more restrictive than necessary, (2) some instructions make no sense or are difficult to understand and (3) some abrasive and coarse labels cause skin irritation. These problems can be avoided by the manufacturers with necessary action. The conditions essential for a clothing care label should always be clearly written out using a universal language or symbol. The selection of soft material for preparing the labels or directly printing the instructions on some inner part of a textile item can avoid the problems of skin irritation.

The use of the Internet in selling various products has grown tremendously. The purchasing of clothing from the Internet poses the risk of size, fit, aesthetics, feel of the fabrics and read information on the care and content labels [382–384].

In addition to the essential information, the manufacturers can include additional information such as environmental labels, guarantees, finishing information and a sweatshop label. Environmental labels provide information on environmentally friendly, ozone-friendly, biodegradability and recycling. Environmental labels can assist consumers in the selection of products with lower impact on the environment. With the guarantees in the clothing, consumers are assured on the quality of the product. The information on textile finish can provide additional information to the consumers. The sweatshop label provides information on the employee's working conditions during the textile production process.

The consumers should also properly understand the meaning of the care instructions before any process. A survey found that many people do not fully understand care label information and select more vigorous cleaning methods than those recommended. Some respondents indicated that they thought bleaching was acceptable, though the instruction warned against it. Similarly, 'line dry' was interpreted incorrectly. Educational programs are therefore necessary to maximize the number of consumers correctly interpreting the labels. Standardizing information on care labels can also minimize misunderstanding. It is essential for the manufacturers to always include the care label with the right care instructions that will be an integral part of the clothing for the useful life of a product. However, several researchers have demonstrated no existence of a direct relationship between information provided and information used.

There are professional wet cleaning businesses that use water and biodegradable soap for washing leather, suede, most tailored woollens, silk and rayon items. The use of tensioning machines and moisture-controlled dryers ensures the fabric retains its original size and shape. These establishments clean the majority of the garments labelled as 'dry clean only' safely to a satisfactory level. Similarly, the garments with a 'dry clean only' label should be cleaned by a professional business in order to avoid any possible damage to the clothing.

Crease-resist garments have been in use for quite some time. The use of resins and cross-linking agents is well established for cotton and its blends. In the case of synthetics, especially polyester, heat-setting techniques have been well-researched. These finishes have to be taken into consideration while preparing care labels.

Soil-resist and anti-microbial finishes are made possible by the use of specialty chemicals [41,385]. Such chemicals have to be retained after washing so as to maintain the desired characteristics. At times, the manufacturer also recommends the topping up of such finishes. Milder washing cycles are sufficient to ensure cleanliness.

Low-water washing (levis) is a new introduction. Herein, modern technology has been used to assist users in long-term care of garments. The concept aims to draw customers based on the environmental benefits.

Self-cleaning garments and fragrant fabrics (microencapsulation) are areas under development. They use intelligent microbes and nanotechnology to impart functionality to fabrics and garments. The care system for such materials will need to be established once they are widely available.

As the use of protective clothing is becoming increasingly important now, and there are several types varying in material and design, the care and maintenance of these items is rather difficult. Several protective clothing items lack standards describing their care and maintenance. Hence, the standards bodies around the globe should come forward to establish standards for the care and maintenance of protective clothing.

Technological developments have helped to prepare environmentally friendly detergents by changing the formulations. The use of enzymes in detergent compositions helps to work at lower temperatures, thereby reducing energy consumption [386]. As these detergents are compact, they reduce packaging, transportation and storage needs resulting in environmental benefits [387].

The future improvement will focus on the replacement of harmful chemicals with bio-based degradable ingredients. Newer detergents are being manufactured with natural-base surfactants such as palm- or coconut oil-based alcohol modified with ethylene oxide. Vegetable oil-based alcohols mostly with 12 carbon atoms foam well in solvent-like water. Methyl esters from palm oil are combined with ethylene oxide to make methyl ester ethoxylates that effectively clean soiled clothes. Alkylpolyethoxide (APE) nonionic detergents prepared from palm- or coconut-based detergent alcohols modified with ethylene oxide are also good cleaning agents.

chapter ten

Conclusions

If a garment does not meet its performance requirements, it fails to meet its business objectives. Manufacturers and retailers will suffer losses because of returns, complaints and reputational damage with their target market. The durability of a garment depends mainly on its care (i.e., severity in laundering, dry cleaning and ironing.) Garment performance can be enhanced by the appropriate selection of fibre, yarn, fabric; combination of production processes and application of finishes. Standard test methods can also be established to check whether the garment meets the performance required for the intended use before it goes to the consumer.

Consumers do not have the experience and technical knowledge to decide which care treatment is suitable for a product. Thus, care labelling is the responsibility of garment makers to help the consumers to maintain the apparel's aesthetic value and durability. The manufacturer is responsible for proper labelling of textile fibre products when they are ready for sale or delivery to the consumer. The importer is responsible for proper labelling of imported textile products. Custom merchants and tailors are responsible for showing properly labelled bolts, samples and swatches to customers. Domestic manufacturers must attach care labels to finished products before they sell them.

For consumer care, symbols make sense when they can understand and follow the instructions. Symbols should provide the same information to everyone without language barriers. Use of symbols allows for smaller and more comfortable care labels, and the symbols are easy to understand. Smaller labels also cost less and this could translate into consumer savings. For manufacturers, care symbols make even more sense. When harmonized with other countries, symbols will allow participation in a global marketplace where symbols will clearly communicate the same information in all countries. Smaller labels cost less to buy or manufacture and also cost less to inventory. Total inventory can be further reduced by eliminating the need for different labels for different countries. Therefore, all the manufacturers should attach care labelling instructions to the garment for the benefit of the consumers and to upkeep their brand.

New developments in fibre technology (such as microfibres, nanofibres and speciality fibres) and finishes extend the analytical aspect of garment manufacture. Similar developments in other areas (such as laundering chemicals and techniques) are necessary to cope with these advanced materials. Products with advanced fibres and finishes will also require the development of new care instructions; hence, the existing care instruction process should be updated accordingly.

References

1. Orzada, B.T., et al., Effect of laundering on fabric drape, bending and shear. *International Journal of Clothing Science and Technology*, 2009. **21**(1): 44–55.
2. Nayak, R. and R. Padhye, The care of apparel products, in *Textiles and Fashion: Materials, Design and Technology*, R. Sinclair, Editor. 2014, Woodhead-Elsevier: Cambridge, UK, pp. 799–822.
3. Reddy, N., A. Salam, and Y. Yang, Effect of structures and concentrations of softeners on the performance properties and durability to laundering of cotton fabrics. *Industrial & Engineering Chemistry Research*, 2008. **47**(8): 2502–2510.
4. Nayak, R. and R. Padhye, Care labelling of clothing, in *Garment Manufacturing Technology*, R. Nayak and R. Padhye, Editors. 2015, Woodhead-Elsevier: Cambridge, UK.
5. Laitala, K. and C. Boks, Sustainable clothing design: Use matters. *Journal of Design Research*, 2012. **10**(1–2): 121–139.
6. Hustvedt, G., Review of laundry energy efficiency studies conducted by the US Department of Energy. *International Journal of Consumer Studies*, 2011. **35**(2): 228–236.
7. ASTM (American Society for Testing and Materials) International, *ASTM D 5489 – 98, Standard Guide for Care Symbols for Care Instructions on Textile Products*. 1998.
8. Ashworth, P., Textile care labelling. *Journal of Consumer Studies & Home Economics*, 1978. **2**(4): 313–322.
9. Mupfumira, I.M. and N. Jinga, *Clothing Care Manual*. 2014, Strategic Book Publishing and Rights Agency; Texas.
10. Yan, R., J. Yurchisin, and K. Watchravesringkan, Use of care labels: Linking need for cognition with consumer confidence and perceived risk. *Journal of Fashion Marketing and Management*, 2008. **12**(4): 532–544.
11. Merwe, D., et al., Consumers' knowledge of textile label information: An exploratory investigation. *International Journal of Consumer Studies*, 2014. **38**(1): 18–24.
12. Sinha, P. and C. Hussey, *Product Labelling for Improved End-of-Life Management*. Product Labelling for Improved End-of-Life Management. 2009.
13. Babel, S. and K. Arora, Knowledge of garment care symbols among women consumers. *Man-Made Textiles in India*, 2008. **51**(3): 37–42.
14. Quaynor, L., M. Nakajima, and M. Takahashi, Dimensional changes in knitted silk and cotton fabrics with laundering. *Textile Research Journal*, 1999. **69**(4): 285–291.

15. Quaynor, L., M. Takahashi, and M. Nakajima, Effects of laundering on the surface properties and dimensional stability of plain knitted fabrics. *Textile Research Journal*, 2000. **70**(1): 28–35.

16. Wilcock, A. and E. Delden, A study of the effects of repeated commercial launderings on the performance of 50/50 polyester/cotton momie cloth. *Journal of Consumer Studies & Home Economics*, 1985. **9**(3): 275–281.

17. Anand, S., et al., Effect of laundering on the dimensional stability and distortion of knitted fabrics. *AUTEX Research Journal*, 2002. **2**(2): 85–100.

18. Higgins, L., et al., Effects of various home laundering practices on the dimensional stability, wrinkling, and other properties of plain woven cotton fabrics. *Textile Research Journal*, 2003. **73**(5): 407.

19. Mackay, C., S.C. Anand, and D.P. Bishop, Effects of laundering on the sensory and mechanical properties of 1 × 1 rib knitwear fabrics part I: Experimental procedures and fabric dimensional properties. *Textile Research Journal*, 1996. **66**(3): 151–157.

20. Perkins, R., G. Drake, and W. Reeves, The effect of laundering variables on the flame retardancy of cotton fabrics. *Journal of the American Oil Chemists' Society*, 1971. **48**(7): 330–333.

21. Wilkinson, P. and R. Hoffman, The effects of wear and laundering on the wrinkling of fabrics. *Textile Research Journal*, 1959. **29**(8): 652–660.

22. Williams, B.L. and P. Horridge, Effects of selected laundering and dry-cleaning pretreatments on the colors of naturally colored cotton. *Family and Consumer Sciences Research Journal*, 1996. **25**(2): 137–158.

23. Baker, M.W. and B.M. Reagan, The effect of dry-cleaning and of laundering with and without fabric softeners on the thermal properties of blankets1. *Textile Research Journal*, 1979. **49**(5): 302–308.

24. Spitz, L., *Soap Technology for the 1990s*. 1990, American Oil Chemists Society; Urbana, IL.

25. Bubl, J.L., Laundering cotton fabric part I: Effects of detergent type and water temperature on soil removal. *Textile Research Journal*, 1970. **40**(7): 637–643.

26. Morris, M.A., Laundering cotton fabric part II: Effect of detergent type and water temperature on appearance, hand, strength, and cost. *Textile Research Journal*, 1970. **40**(7): 644–649.

27. Morton, G. and L. Thomas, Interactions of fiber, finish, and wash conditions in laundering cotton-containing fabrics. *Textile Research Journal*, 1983. **53**(3): 165.

28. Webb, J.J. and S.K. Obendorf, Detergency study: Comparison of the distribution of natural residual soils after laundering with a variety of detergent products. *Textile Research Journal*, 1987. **57**(11): 640–646.

29. Fort, T., H. Billica, and T. Grindstaff, Studies of soiling and detergency part II: Detergency experiments with model fatty soils. *Textile Research Journal*, 1966. **36**(2): 99–112.

30. Nagarajan, M. and H. Paine, Water hardness control by detergent builders. *Journal of the American Oil Chemists Society*, 1984. **61**(9): 1475–1478.

31. Laughlin, J. and R.E. Gold, Water hardness, detergent type, and prewash product use as factors affecting methyl parathion residue retained in protective apparel fabrics. *Clothing and Textiles Research Journal*, 1990. **8**(4): 61–67.

32. Arai, H., Study of detergency. I. Effect of the concentration and the kind of detergent in hard water. *Journal of the American Oil Chemists Society*, 1966. **43**(5): 312–314.

33. Sherman, P.O., S. Smith, and B. Johannessen, Textile characteristics affecting the release of soil during laundering. Part II: Fluorochemical soil-release textile finishes. *Textile Research Journal*, 1969. **39**(5): 449–459.
34. Cutler, G., *Detergency: Theory and Technology*. Vol. 20. 1986, CRC Press; Boca Raton, FL.
35. Bourne, M. and W. Jennings, Kinetic studies of detergency. II. Effect of age, temperature, and cleaning time on rates of soil removal. *Journal of the American Oil Chemists' Society*, 1963. **40**(10): 523–530.
36. Kissa, E., Kinetics and mechanisms of detergency part I: Liquid hydrophobic (oily) soils. *Textile Research Journal*, 1975. **45**(10): 736–741.
37. Compton, J. and W. Hart, Soiling and soil retention in textile fibers. Cotton fiber-grease-free carbon black system. *Industrial & Engineering Chemistry*, 1951. **43**(7): 1564–1569.
38. Goyal, A. and R. Nayak, Some quality aspects of cotton ring-and compact-spun yarn. *Man-Made Textiles in India*, 2006. **49**(7): 271–273.
39. Huh, Y., Y.R. Kim, and W. Oxenham, Analyzing structural and physical properties of ring, rotor, and friction spun yarns. *Textile Research Journal*, 2002. **72**(2): 156–163.
40. Behera, B.K. and P. Hari, *Woven Textile Structure: Theory and Applications*. 2010, Woodhead-Elsevier; Cambridge, UK.
41. Nayak, R. and R. Padhye, Antimicrobial finishes for textiles, in *Functional Finishes for Textiles: Improving Comfort, Performance and Protection*, R. Paul, Editor. 2014, Woodhead-Elsevier; Cambridge, UK.
42. Nayak, R., et al., Evaluation of functional finishes: An overview. *Man-Made Textiles in India*, 2008. **51**(4), 130–135.
43. Kissa, E., Kinetics of oily soil release. *Textile Research Journal*, 1971. **41**(9): 760–767.
44. Kissa, E., Mechanisms of soil release. *Textile Research Journal*, 1981. **51**(8): 508–513.
45. Smith, S. and P.O. Sherman, Textile characteristics affecting the release of soil during laundering part I: A review and theoretical consideration of the effects of fiber surface energy and fabric construction on soil release. *Textile Research Journal*, 1969. **39**(5): 441–449.
46. Bajpai, D. and V. Tyagi, Laundry detergents: An overview. *Journal of Oleo Science*, 2007. **56**(7): 327–340.
47. Lowe, E., W. Adair, and E. Johnston, Soaps and detergents—The inorganic components. *Journal of the American Oil Chemists' Society*, 1978. **55**(1): 32–35.
48. Easley, C., et al., Detergents and water temperature as factors in methyl parathion removal from denim fabrics. *Bulletin of Environmental Contamination and Toxicology*, 1982. **28**(2): 239–244.
49. Sams, P. Clothes care-sending the right signals. in *International Appliance Technical Conference*, Columbus, OH, 27th March. 2001.
50. Shove, E., Converging conventions of comfort, cleanliness and convenience. *Journal of Consumer Policy*, 2003. **26**(4): 395–418.
51. Purchase, M.E., C.K. Berning, and A.L. Lyng, The cost of washing clothes: Sources of variation. *Journal of Consumer Studies & Home Economics*, 1982. **6**(4): 301–317.
52. Cameron, B.A., Detergent considerations for consumers: Laundering in hard water–how much extra detergent is required. *The Journal of Extension*, 2011. **49**: 4RIB6.

53. Arild, A.-H., et al., *An Investigation of Domestic Laundry in Europe: Habits, Hygiene and Technical Performance*. 2003: SIFO, Statens Institutt for Forbruksforskning Stensberggata, Norway.

54. Pakula, C. and R. Stamminger, Electricity and water consumption for laundry washing by washing machine worldwide. *Energy Efficiency*, 2010. **3**(4): 365–382.

55. Stamminger, R. Consumer real life behaviour compared to standard in washing and dishwashing, in *Proceedings of the WFK 44th International Detergency Conference*. 2009, Duesseldorf, Germany.

56. Allwood, J., et al., *Well Dressed? The Present and Future Sustainability of Clothing and Textiles in the United Kingdom*. 2006, University of Cambridge, Institute for Manufacturing: Cambridge.

57. Shove, E., *Comfort, Cleanliness and Convenience: The Social Organization of Normality*. 2003, Berg; Oxford.

58. Phillips, D., et al., Effect of liquor ratio on the shade change and cross-staining observed in the ISO 105-C08 test†. *Coloration Technology*, 2003. **119**(3): 177–181.

59. Bishop, D., Physical and chemical effects of domestic laundering processes, in *Chemistry of the Textiles Industry*. C.M. Carr, Editor. 1995, Springer: Singapore, pp. 125–172.

60. Nayak, R. and R. Padhye, Care of apparel products, in *Textiles and Fashion: Materials, Design and Technology*, R. Sinclair, Editor. 2014, Elsevier; Cambridge, UK, pp. 799–822.

61. Moody, R.P. and H.I. Maibach, Skin decontamination: Importance of the wash-in effect. *Food and Chemical Toxicology*, 2006. **44**(11): 1783–1788.

62. Simion, F.A., et al., Self-perceived sensory responses to soap and synthetic detergent bars correlate with clinical signs of irritation. *Journal of the American Academy of Dermatology*, 1995. **32**(2): 205–211.

63. Slotosch, C.M., G. Kampf, and H. Löffler, Effects of disinfectants and detergents on skin irritation. *Contact Dermatitis*, 2007. **57**(4): 235–241.

64. Rodriguez, C., et al., Skin effects associated with wearing fabrics washed with commercial laundry detergents. *Journal of Toxicology: Cutaneous and Ocular Toxicology*, 1994. **13**(1): 39–45.

65. Toshima, Y., et al., Observation of everyday hand-washing behavior of Japanese, and effects of antibacterial soap. *International Journal of Food Microbiology*, 2001. **68**(1): 83–91.

66. Ojajärvi, J., Effectiveness of hand washing and disinfection methods in removing transient bacteria after patient nursing. *Journal of Hygiene*, 1980. **85**(02): 193–203.

67. Chun-Yoon, J. and C.R. Jasper, Consumer preferences for size description systems of men's and women's apparel. *Journal of Consumer Affairs*, 1995. **29**(2): 429–441.

68. Collier, A.M., *A Handbook of Textiles*. 1970: Pergamon Press, Oxford, UK.

69. Larson, E. and E. Lusk, Evaluating handwashing technique. *Journal of Advanced Nursing*, 2006. **53**(1): 46–50.

70. Puchta, R., Cationic surfactants in laundry detergents and laundry aftertreatment aids. *Journal of the American Oil Chemists' Society*, 1984. **61**(2): 367–376.

71. Zoller, U. and P. Sosis, *Handbook of Detergents, Part F: Production*. Vol. 142. 2008, CRC Press; Boca Raton, FL.

72. Shaeiwitz, J., et al., The mechanism of solubilization in detergent solutions. *Journal of Colloid and Interface Science*, 1981. **84**(1): 47–56.
73. Bertleff, W., et al., Aspects of polymer use in detergents. *Journal of Surfactants and Detergents*, 1998. **1**(3): 419–424.
74. Aboul-Kassim, T.A. and B.R. Simoneit, Detergents: A review of the nature, chemistry, and behavior in the aquatic environment. Part I. Chemical composition and analytical techniques. *Critical Reviews in Environmental Science and Technology*, 1993. **23**(4): 325–376.
75. Broze, G., *Handbook of Detergents: Properties*. 1999, CRC Press; Boca Raton, FL.
76. Parris, N., J. Weil, and W. Linfield, Soap based detergent formulations. V. Amphoteric lime soap dispersing agents. *Journal of the American Oil Chemists Society*, 1973. **50**(12): 509–512.
77. Sheets, W.D. and G.W. Malaney, Synthetic detergents and the BOD test. *Sewage and Industrial Wastes*, 1956. **28**(1): 10–17.
78. Van Os, N.M., J.R. Haak, and L.A.M. Rupert, *Physico-Chemical Properties of Selected Anionic, Cationic and Nonionic Surfactants*. 2012, Elsevier: Amsterdam, Netherlands.
79. Scheibel, J.J., The evolution of anionic surfactant technology to meet the requirements of the laundry detergent industry. *Journal of Surfactants and Detergents*, 2004. **7**(4): 319–328.
80. Vora, B., et al., Recent advances in the production of detergent olefins and linear alkylbenzenes. *Tenside, Surfactants, Detergents*, 1991. **28**(4): 287–294.
81. Toedt, J., D. Koza, and K. Van Cleef-Toedt, *Chemical Composition of Everyday Products*. 2005, Greenwood Publishing Group: Westport, CT.
82. Yangxin, Y., Z. Jin, and A.E. Bayly, Development of surfactants and builders in detergent formulations. *Chinese Journal of Chemical Engineering*, 2008. **16**(4): 517–527.
83. Lewis, M., Chronic toxicities of surfactants and detergent builders to algae: A review and risk assessment. *Ecotoxicology and Environmental Safety*, 1990. **20**(2): 123–140.
84. Rutkowski, B.J., Recent changes in laundry detergents, in *Detergency, Theory and Test Methods*, 1981. Marcel Dekker, New YorkEditor: F. L. Howard.
85. van Ee, J.H. and O. Misset, *Enzymes in Detergency*. Vol. 69. 1997, CRC Press: Boca Raton, FL.
86. Findley, W.R., Fluorescent whitening agents for modern detergents. *Journal of the American Oil Chemists' Society*, 1988. **65**(4): 679–683.
87. Olsen, H.S. and P. Falholt, The role of enzymes in modern detergency. *Journal of Surfactants and Detergents*, 1998. **1**(4): 555–567.
88. Lyle, D., *Performance of Textiles*. 1977: John Wiley & Sons, Hoboken, New Jersey.
89. Illman, J., T. Albin, and H. Stupel, Studies on replacement of phosphate builders in laundry detergents using radiolabeled soils. *Journal of the American Oil Chemists' Society*, 1972. **49**(4): 217–221.
90. Hollingsworth, M., Role of detergent builders in fabric washing formulations. *Journal of the American Oil Chemists' Society*, 1978. **55**(1): 49–51.
91. Jungermann, E. and H. Silberman, Carbonate and phosphate detergent builders: Their impact on the environment. *Journal of the American Oil Chemists' Society*, 1972. **49**(8): 481–484.

92. Kim, J.O., B.F. Smith, and S.M. Spivak, Comparative study of phosphate and non-phosphate detergents. *Journal of Consumer Studies and Home Economics,* 1987. **11**(3): 219–235.

93. Greek, B., Sales of detergents growing despite recession. *Chemical and Engineering News,* 1991. **69**(4): 25–52.

94. Lai, K.-Y., *Liquid Detergents.* Vol. 129. 2005: CRC Press, Boca Raton, Florida.

95. Association, S.a.D., *Types of Laundry Products. Laundering Facts.* 1991, Soap and Detergent Association: New York. p. 1.

96. Lin, J. and M. Iyer, Cold or hot wash: Technological choices, cultural change, and their impact on clothes-washing energy use in China. *Energy Policy,* 2007. **35**(5): 3046–3052.

97. Davis, S. and P. Ainsworth, The disinfectant action of low-temperature laundering. *Journal of Consumer Studies & Home Economics,* 1989. **13**(1): 61–66.

98. Cameron, B.A., D.M. Brown, and S.S. Meyer, A survey of commercial laundry detergents-How effective are they? Part II: Liquids. *Journal of Consumer Studies and Home Economics,* 1993. **17**(3): 267–273.

99. Vaughn, T., et al., New developments in army sea water laundering. *Industrial & Engineering Chemistry,* 1949. **41**(1): 112–119.

100. Cameron, B.A., Laundering in cold water: Detergent considerations for consumers. *Family and Consumer Sciences Research Journal,* 2007. **36**(2): 151–162.

101. Laitala, K., C. Boks, and I.G. Klepp, Potential for environmental improvements in laundering. *International Journal of Consumer Studies,* 2011. **35**(2): 254–264.

102. Coons, D., Bleach: Facts, fantasy, and fundamentals. *Journal of the American Oil Chemists' Society,* 1978. **55**(1): 104–108.

103. Jordan, W.E., D.V. Jones, and M. Klein, Antiviral effectiveness of chlorine bleach in household laundry use. *American Journal of Diseases of Children,* 1969. **117**(3): 313–316.

104. Milne, N.J., Oxygen bleaching systems in domestic laundry. *Journal of Surfactants and Detergents,* 1998. **1**(2): 253–261.

105. Belkin, N.L., Aseptics and aesthetics of chlorine bleach: Can its use in laundering be safely abandoned? *American Journal of Infection Control,* 1998. **26**(2): 149–151.

106. Racioppi, F., et al., Household bleaches based on sodium hypochlorite: Review of acute toxicology and poison control center experience. *Food and Chemical Toxicology,* 1994. **32**(9): 845–861.

107. Ainsworth, P. and J. Fletcher, A comparison of the disinfectant action of a powder and liquid detergent during low-temperature laundering. *Journal of Consumer Studies & Home Economics,* 1993. **17**(1): 67–73.

108. Kennedy, J.A., J. Bek, and D. Griffin, *Selection and Use of Disinfectants.* 2000: Cooperative Extension, Institute of Agriculture and Natural Resources, University of Nebraska-Lincoln, Lincoln, Nebraska.

109. Lawrence, C.A., *Surface-Active Quaternary Ammonium Germicides.* 2013: Academic Press, Cambridge, Massachusetts.

110. Avinc, O., et al., Effects of softeners and laundering on the handle of knitted PLA filament fabrics. *Fibers and Polymers,* 2010. **11**(6): 924–931.

111. Levinson, M.I., Rinse-added fabric softener technology at the close of the twentieth century. *Journal of Surfactants and Detergents,* 1999. **2**(2): 223–235.

112. Baumert, K.J., J. Penney, and P. Cox Crews, Influence of household fabric softeners on properties of selected woven fabrics. *Textile Chemist and Colorist,* 1996. **28**(4): 36–43.

113. Parthiban, M. and R. Kumar, Effect of fabric softener on thermal comfort of cotton and polyester fabrics. *Indian Journal of Fibre & Textile Research*, 2007. **32**(4): 446–452.

114. Obendorf, S., et al., Starch as a renewable finish to improve the pesticide-protective properties of conventional workclothes. *Archives of Environmental Contamination and Toxicology*, 1991. **21**(1): 10–16.

115. Ko, L. and S. Obendorf, Effect of starch on reducing the retention of methyl parathion by cotton and polyester fabrics in agricultural protective clothing. *Journal of Environmental Science & Health Part B*, 1997. **32**(2): 283–294.

116. Haghayegh, G. and R. Schoenlechner, Physically modified starches: A review. *Journal of Food Agric Environment*, 2011. **9**: 27–29.

117. Von Grote, J., et al., Assessing occupational exposure to perchloroethylene in dry cleaning. *Journal of Occupational and Environmental Hygiene*, 2006. **3**(11): 606–619.

118. Kovacs, D.C., B. Fischhoff, and M.J. Small, Perceptions of PCE use by dry cleaners and dry cleaning customers. *Journal of Risk Research*, 2001. **4**(4): 353–375.

119. Räisänen, J., R. Niemelä, and C. Rosenberg, Tetrachloroethylene emissions and exposure in dry cleaning. *Journal of the Air & Waste Management Association*, 2001. **51**(12): 1671–1675.

120. Tichenor, B.A., et al., Emissions of perchloroethylene from dry cleaned fabrics. *Atmospheric Environment. Part A. General Topics*, 1990. **24**(5): 1219–1229.

121. Solet, D., T.G. Robins, and C. Sampaio, Perchloroethylene exposure assessment among dry cleaning workers. *The American Industrial Hygiene Association Journal*, 1990. **51**(10): 566–574.

122. Altmann, L., et al., Neurobehavioral and neurophysiological outcome of chronic low-level tetrachloroethene exposure measured in neighborhoods of dry cleaning shops. *Environmental Research*, 1995. **69**(2): 83–89.

123. Kamrin, M.A. and A. Director, *The Scientific Facts about the Dry-Cleaning Chemical Perc*. 2001: American Council on Science and Health, New York.

124. McGregor, D.B., E. Heseltine, and H. Møller, Dry cleaning, some solvents used in dry cleaning and other industrial chemicals. *Scandinavian Journal of Work, Environment & Health*, 1995: **21**(4): 310–312.

125. Blair, A., P. Decoufle, and D. Grauman, Causes of death among laundry and dry cleaning workers. *American Journal of Public Health*, 1979. **69**(5): 508–511.

126. Katz, R.M. and D. Jowett, Female laundry and dry cleaning workers in Wisconsin: A mortality analysis. *American Journal of Public Health*, 1981. **71**(3): 305–307.

127. DeSimone, J.M., Practical approaches to green solvents. *Science*, 2002. **297**(5582): 799–803.

128. Eastoe, J., et al., Micellization of hydrocarbon surfactants in supercritical carbon dioxide. *Journal of the American Chemical Society*, 2001. **123**(5): 988–989.

129. Mutti, A., et al., Nephropathies and exposure to perchloroethylene in dry-cleaners. *The Lancet*, 1992. **340**(8813): 189–193.

130. Van Roosmalen, M., G. Woerlee, and G. Witkamp, Dry-cleaning with high pressure carbon dioxide: The influence of process conditions and various co-solvents (alcohols) on cleaning results. *The Journal of Supercritical Fluids*, 2003. **27**(3): 337–344.

131. Van Roosmalen, M., G. Woerlee, and G. Witkamp, Surfactants for particulate soil removal in dry-cleaning with high-pressure carbon dioxide. *The Journal of Supercritical Fluids*, 2004. **30**(1): 97–109.

132. Goddard, E., Polymer/surfactant interaction—its relevance to detergent systems. *Journal of the American Oil Chemists' Society,* 1994. **71**(1): 1–16.
133. Dabestani, A., Dry cleaning surfactants. *Surfactant Science Series,* 2001: 123(38): 9492–9492.
134. Heyns, M., et al., Advanced wet and dry cleaning coming together for next generation. *Solid State Technol,* 1999. **42**(3): 37–44.
135. Wallace, L.A., et al., The influence of personal activities on exposure to volatile organic compounds. *Environmental Research,* 1989. **50**(1): 37–55.
136. Tokino, S., et al., Laundering shrinkage of wool fabric treated with low-temperature plasmas under atmospheric pressure. *Journal of the Society of Dyers and Colourists,* 1993. **109**(10): 334–335.
137. Onal, L. and C. Candan, Contribution of fabric characteristics and laundering to shrinkage of weft knitted fabrics. *Textile Research Journal,* 2003. **73**(3): 187–191.
138. Altham, W., Benchmarking to trigger cleaner production in small businesses: Drycleaning case study. *Journal of Cleaner Production,* 2007. **15**(8): 798–813.
139. Chen, H.-L. and L.D. Burns, Environmental analysis of textile products. *Clothing and Textiles Research Journal,* 2006. **24**(3): 248–261.
140. Ludwig, H.R., et al., Worker exposure to perchloroethylene in the commercial dry cleaning industry. *The American Industrial Hygiene Association Journal,* 1983. **44**(8): 600–605.
141. Lohman, J.H., A history of dry cleaners and sources of solvent releases from dry cleaning equipment. *Environmental Forensics,* 2002. **3**(1): 35–58.
142. Ott, W.R. and J.W. Roberts, Everyday exposure to toxic pollutants. *Scientific American,* 1998. **278**(2): 72–77.
143. Ameen, A. and S. Bari, Investigation into the effectiveness of heat pump assisted clothes dryer for humid tropics. *Energy Conversion and Management,* 2004. **45**(9): 1397–1405.
144. Laing, R.M., et al., Determining the drying time of apparel fabrics. *Textile Research Journal,* 2007. **77**(8): 583–590.
145. Madsen, J.D., Biomass techniques for monitoring and assessing control of aquatic vegetation. *Lake and Reservoir Management,* 1993. **7**(2): 141–154.
146. Han, H. and S. Deng, A study on residential clothes drying using waste heat rejected from a split-type room air conditioner (RAC). *Drying Technology,* 2003. **21**(8): 1471–1490.
147. Bansal, P., J. Braun, and E. Groll, Improving the energy efficiency of conventional tumbler clothes drying systems. *International Journal of Energy Research,* 2001. **25**(15): 1315–1332.
148. Ng, A.B. and S. Deng, A new termination control method for a clothes drying process in a clothes dryer. *Applied Energy,* 2008. **85**(9): 818–829.
149. Yadav, V. and C. Moon, Fabric-drying process in domestic dryers. *Applied Energy,* 2008. **85**(2): 143–158.
150. Berghel, J., L. Brunzell, and P. Bengtsson. Performance analysis of a tumble dryer, in *Proceedings of the 14th International Dryer Symposium (IDS 2004), B.* 2004, Saõ Paulo, Brazil.
151. Moon, H.J. and Y.R. Yoon, Investigation of physical characteristics of houses and occupants' behavioural factors for mould infestation in residential buildings. *Indoor and Built Environment,* 2010. **19**(1): 57–64.

152. Rousseau, M., et al. Characterization of indoor hygrothermal conditions in houses in different northern climates, in *Thermal Performance of Exterior Envelopes of Whole Buildings X International Conference.* 2007, Florida.

153. Ferguson, M., Introduction: The apparel industry, in *Garment Manufacturing Technology*, R. Nayak and R. Padhye, Editors. 2015, Woodhead: Cambridge, UK, pp. 387–404.

154. Zhibin, X. and S. Baoshou. Research of ironing product by saturated steam thermal energy, in *Measuring Technology and Mechatronics Automation (ICMTMA), 2010 International Conference on IEEE.* 2010, Wuhan Hubei, China.

155. Ahuja, P., R. Kaur, and P. Sandhu, Comparative evaluation of selected irons on basis of ergonomic costs of ironing clothes on workers body. *Journal of Dairying Foods & Home Sciences*, 2002. **21**(2): 136–139.

156. Hatch, K.L., American standards for UV-protective textiles, in *Cancers of the Skin*. R. Dummer, F.O. Nestle, G. Burg, Editors. 2002, Springer: Singapore,. pp. 42–47.

157. Wang, L., et al., Recent trends in ballistic protection. *Textiles and Light Industrial Science and Technology*, 2014. **3**: 37–47.

158. Nayak, R., S. Houshyar, and R. Padhye, Recent trends and future scope in the protection and comfort of fire-fighters' personal protective clothing. *Fire Science Reviews*, 2014. **3**(1): 1–19.

159. Swinker, M.E. and J.D. Hines, Understanding consumers' perception of clothing quality: A multidimensional approach. *International Journal of Consumer Studies*, 2006. **30**(2): 218–223.

160. Palomo-Lovinski, N., Extensible dress the future of digital clothing. *Clothing and Textiles Research Journal*, 2008. **26**(2): 119–130.

161. Kay Obendorf, S., Microscopy to define soil, fabric and detergent formulation characteristics that affect detergency: A review. *AATCC Review*, 2004. **4**(1): 17–23.

162. Crown, E.M., A. Feng, and X. Xu, How clean is clean enough? Maintaining thermal protective clothing under field conditions in the oil and gas sector. *International Journal of Occupational Safety and Ergonomics*, 2004. **10**(3): 247–254.

163. Rezazadeh, M. and D.A. Torvi, Assessment of factors affecting the continuing performance of firefighters' protective clothing: A literature review. *Fire Technology*, 2011. **47**(3): 565–599.

164. Järvi, P. and A. Paloviita, Product-related information for sustainable use of laundry detergents in Finnish households. *Journal of Cleaner Production*, 2007. **15**(7): 681–689.

165. Chin, J., et al., Effect of artificial perspiration and cleaning chemicals on the mechanical and chemical properties of ballistic materials. *Journal of Applied Polymer Science*, 2009. **113**(1): 567–584.

166. NIJ, *Body Armor User Guide*, National Institute of Justice: Washington, DC, 2000.

167. Fahl, K., Body armor technology: How to properly fit and maintain soft body armor. *Law Enforcement Technology*, 2007. **34**(2): 92–94.

168. Nayak, R., L. Wang, and R. Padhye, Electronic textiles for military personnel, in *Electronic Textiles: Smart Fabrics and Wearable Technology*, T. Dias, Editor. 2015, Woodhead-Elsevier: Cambridge, UK.

169. Helliker, M., et al., Effect of domestic laundering on the fragment protective performance of fabrics used in personal protection. *Textile Research Journal*, 84(12): 1298: 1306, 2014: 0040517513512400.

170. Taylor, M.A., *Technology of Textile Properties: An Introduction*. 1990, Forbes Forbes Publications Ltd: New York.

171. Lavy, T., J. Mattice, and R. Flynn, Field studies monitoring worker exposure to pesticides. *Pesticide Formulations and Application Systems*, 1983: 60–74.

172. Gotoh, K., Investigation of optimum liquid for textile washing using artificially soiled fabrics. *Textile Research Journal*, 2009 80(6): 548–556.

173. Laughlin, J., K. Newburn, and R.E. Gold, Pyrethroid insecticides and formulations as factors in residues remaining in apparel fabrics after laundering. *Bulletin of Environmental Contamination and Toxicology*, 1991. 47(3): 355–361.

174. Hild, D.N., J.M. Laughlin, and R.E. Gold, Laundry parameters as factors in lowering methyl parathion residue in cotton/polyester fabrics. *Archives of Environmental Contamination and Toxicology*, 1989. 18(6): 908–914.

175. Keaschall, J.L., J.M. Laughlin, and R.E. Gold, Effect of laundering procedures and functional finishes on removal of insecticides selected from three chemical classes. *Performance of Protective Clothing, ASTM STP*, 1986. 900: 162–176.

176. Nelson, C., et al., Laundering as decontamination of apparel fabrics: Residues of pesticides from six chemical classes. *Archives of Environmental Contamination and Toxicology*, 1992. 23(1): 85–90.

177. Nayak, R., R. Padhye, and L. Wang, How to dress at work, in *Management and Leadership: A Guide for Clinical Professionals*, S. Patole, Editor. 2014, Springer International Publishing AG: Singapore.

178. Bruning, L.M., The bloodborne pathogens final rule: Understanding the regulation. *AORN Journal*, 1993. 57(2): 437–466.

179. Yokoe, D.S., et al., A compendium of strategies to prevent healthcare-associated infections in acute care hospitals. *Infection Control and Hospital Epidemiology*, 2008. 29(S1): S12–S21.

180. Belkin, N.L., Home laundering of soiled surgical scrubs: Surgical site infections and the home environment. *American Journal of Infection Control*, 2001. 29(1): 58–64.

181. Doty, K.C. and E. Easter, An analysis of the care and maintenance of performance textiles and effects of care on performance. *AATCC Review*, 2007. 9(5): 37–42.

182. Shaw, A., Selection of flame resistant protective clothing, in *Handbook of Fire Resistant Textiles*, F. Selcen Kilinc, Editor. 2013, Woodhead-Elsevier: Cambridge, UK, p. 351.

183. Stanford, D.G., K.E. Georgouras, and M.T. Pailthorpe, The effect of laundering on the sun protection afforded by a summer weight garment. *Journal of the European Academy of Dermatology and Venereology*, 1995. 5(1): 28–30.

184. Williams, J.T., *Textiles for Cold Weather Apparel*. 2009, Woodhead-Elsevier: Cambridge; UK.

185. Wentz, M., Textile care technology spectra and care labelling issues, in *Apparel Care and Environment: Alternative Technologies and Labelling (EPA 744_R-96-002)*, L. Reppert and L. Speare, Editors. 1996, Eastern Research Group, Inc.: Washington, DC, pp. 83–85.

186. McQueen, R.H., et al., Odor intensity in apparel fabrics and the link with bacterial populations. *Textile Research Journal*, 2007. 77(7): 449–456.

187. Kuklane, K., D. Gavhed, and I. Holmér, Effect of the number, thickness and washing of socks on the thermal insulation of feet. *Ergonomics of Protective Clothing*, NOKOBETEF 6 and 1st European Conference on Protective Clothing, 2000(8): 175–178 Stockholm, Sweden.

188. Kuklane, K., I. Holmer, and G. Giesbrecht, Change of footwear insulation at various sweating rates. *Applied Human Science*, 1999. **18**(5): 161–168.

189. Card, A., M. Moore, and M. Ankeny, Garment washed jeans: Impact of launderings on physical properties. *International Journal of Clothing Science and Technology*, 2006. **18**(1): 43–52.

190. Bal, N., et al., Digital printing of enzymes on textile substrates as functional materials. *Journal of Fiber Bioengineering and Informatics*, 2014. **7**(4): 595–602.

191. Barnett, R. and K. Slater, The progressive deterioration of textile materials part V: The effects of weathering on fabric durability. *Journal of the Textile Institute*, 1991. **82**(4): 417–425.

192. Arunyadej, S., et al., An investigation into the effect of laundering on the repellency behaviour of a fluorochemical-treated cotton fabric. *The Journal of The Textile Institute*, 1998. **89**(4): 696–702.

193. Mukhopadhyay, A., M. Sikka, and A. Karmakar, Impact of laundering on the seam tensile properties of suiting fabric. *International Journal of Clothing Science and Technology*, 2004. **16**(4): 394–403.

194. Hurren, C., P. Cookson, and X. Wang, The effects of ultrasonic agitation in laundering on the properties of wool fabrics. *Ultrasonics Sonochemistry*, 2008. **15**(6): 1069–1074.

195. Morris, M. and H. Prato, Edge abrasion of durable-press cotton fabric during laundering with phosphate-and carbonate-built detergents. *Textile Research Journal*, 1975. **45**(5): 395–401.

196. Hurren, A., A. Wilcock, and K. Slater, The effects of laundering and abrasion on the tensile strength of chemically treated cotton print cloth. *Journal of the Textile Institute*, 1985. **76**(4): 285–288.

197. Neelakantan, P. and H. Mehta, Wear life of easy-care cotton fabrics. *Textile Research Journal*, 1981. **51**(10): 665.

198. Lau, L., et al., Effects of repeated laundering on the performance of garments with wrinkle-free treatment. *Textile Research Journal*, 2002. **72**(10): 931.

199. Chippindale, P., Wear, abrasion, and laundering of cotton fabrics. *Journal of the Textile Institute Transactions*, 1963. **54**(11): 445–463.

200. Murdison, M. and J. Roberts, A study of the effects of laundering and storage on cotton cloth. *Journal of the Textile Institute Transactions*, 1949. **40**(8): 505–518.

201. FiJan, S., et al., The influence of industrial laundering of hospital textiles on the properties of cotton fabrics. *Textile Research Journal*, 2007. **77**(4): 247.

202. Saville, B., *Physical Testing of Textiles*. 1999, CRC Press: Boca Raton, FL.

203. Nayak, R., et al., Sewability of air-jet textured sewing threads in denim. *Journal of Textile and Apparel, Technology and Management*, 2013. **8**(1), 1-11.

204. Nayak, R., R. Padhye, and D.P. Gon, Sewing performance of stretch denim. *Journal of Textile and Apparel, Technology and Management*, 2010. **6**(3), 1–9.

205. Behera, B.K., Role of fabric properties in the clothing-manufacturing process, in *Garment Manufacturing Technology*, R. Nayak and R. Padhye, Editors. 2015, Woodhead Publishing: Cambridge, UK, pp. 59–80.

206. Carvalho, M., et al., Sewing-room problems and solutions, in *Garment Manufacturing Technology*, R. Nayak and R. Padhye, Editors. 2015, Woodhead Publishing: Cambridge, UK, pp. 317–334.

207. Nawaz, N. and R. Nayak, Seamless garments, in *Garment Manufacturing Technology*, R. Nayak and R. Padhye, Editors. 2015, Woodhead Publishing: Cambridge, UK, pp. 373–383.

208. Mccloskey, S.G. and J.M. Jump, Bio-polishing of polyester and polyester/cotton fabric. *Textile Research Journal*, 2005. **75**(6): 480–484.

209. Beltran, R., L. Wang, and X. Wang, A controlled experiment on yarn hairiness and fabric pilling. *Textile Research Journal*, 2007. **77**(3): 179–183.

210. Latifi, M. and H. Kim, Characterizing fabric pilling due to fabric-to-fabric abrasion. *Textile Research Journal*, 2001. **71**(7): 640–644.

211. Elder, H., Wear of Textiles. *Journal of Consumer Studies & Home Economics*, 1978. **2**(1): 1–13.

212. Carvalho, M., et al. Study of pilling formation in wool and blended fabrics using the optical profile analysis, in *Materials Science Forum*. 2004: Trans Tech Publ, Zurich, Switzerland.

213. Booth, J., *Principles of Textile Testing*. 1979 ed. 1968, J W Arrowsmith Ltd: England.

214. Chiweshe, A. and P. Crews, Influence of household fabric softeners and laundry enzymes on pilling and breaking strength. *Textile Chemist and Colorist and American Dyestuff Reporter*, 2000. **32**(9): 41–47.

215. Clegg, G., A microscopic examination of worn textiles. *Journal of the Textile Institute*, 1949. **40**: T449.

216. Backer, S. and S. Tanenhaus, The relationship between the structural geometry of a textile fabric and its physical properties. *Textile Research Journal*, 1951. **21**(9): 635.

217. Morris, M. and H. Prato, Fabric damage during laundering. *California Agriculture*, 1976. **30**(12): 9.

218. DeGruy, I., et al., Microscopical observations of abrasion phenomena in cotton. *Textile Research Journal*, 1962. **32**(11): 873.

219. Gagliardi, D. and A. Wehner, Influence of swelling and mowosubstitution on the strength of cross-linked cotton. *Textile Research Journal*, 1967. **37**(2): 118.

220. Heap, S., Liquid ammonia treatment of cotton fabrics, especially as a pretreatment for easy-care finishing. *Journal of the Textile Institute*, 1978. **16**: 387–390.

221. Reeves, W., Some effects of the nature of cross links on the properties of cotton fabrics. *Journal of the Textile Institute*, 1962: 22–34.

222. Was-Gubala, J., The kinetics of colour change in textiles and fibres treated with detergent solutions: Part I—Colour perception and fluorescence microscopy analysis. *Science & Justice*, 2009. **49**(3): 165–169.

223. Phillips, D., et al., Development of a test to predict colour fading of cotton fabrics after multi-cycle laundering with a bleach-containing domestic detergent. *Journal of the Society of Dyers and Colourists*, 1996. **112**(10): 287–293.

224. Wylie, M., E. Crown, and M. Morris, Consumer reaction to color performance in textiles. *Home Economics Research Journal*, 1977. **5**(3): 167–175.

225. Mangut, M., B. Becerir, and H. Alpay, Effects of repeated home launderings and non-durable press on the colour properties of plain woven polyester fabric. *Indian Journal of Fibre & Textile Research*, 2008. **33**(2): 80–87.

226. Min, L., Z. Xiaoli, and C. Shuilin, Enhancing the wash fastness of dyeings by a sol–gel process. Part 1; Direct dyes on cotton. *Coloration Technology*, 2003. **119**(5): 297–300.

227. El-Shishtawy, R. and S. Nassar, Cationic pretreatment of cotton fabric for anionic dye and pigment printing with better fastness properties. *Coloration Technology*, 2002. **118**(3): 115–120.
228. Schumacher, K., E. Heine, and H. Höcker, Extremozymes for improving wool properties. *Journal of Biotechnology*, 2001. **89**(2–3): 281–288.
229. Xie, K., A. Hou, and Y. Zhang, New polymer materials based on silicone acrylic copolymer to improve fastness properties of reactive dyes on cotton fabrics. *Journal of Applied Polymer Science*, 2006. **100**(1): 720–725.
230. Topalbekiroglu, M. and H.K. Kaynak, The effect of weave type on dimensional stability of woven fabrics. *International Journal of Clothing Science and Technology*, 2008. **20**(5): 281–288.
231. Higgins, L., et al., Effect of tumble-drying on selected properties of knitted and woven cotton fabrics: Part I: Experimental overview and the relationship between temperature setting, time in the dryer and moisture content. *Journal of the Textile Institute*, 2003. **94**(1–2): 119–128.
232. Lin, J.-J., Prediction of yarn shrinkage using neural nets. *Textile Research Journal*, 2007. **77**(5): 336–342.
233. Higgins, L., et al., Effects of various home laundering practices on the dimensional stability, wrinkling, and other properties of plain woven cotton fabrics part II: Effect of rinse cycle softener and drying method and of tumble sheet softener and tumble drying time. *Textile Research Journal*, 2003. **73**(5): 407–420.
234. Das, D. and R. Thakur, Taguchi analysis of fabric shrinkage. *Fibers and Polymers*, 2013. **14**(3): 482–487.
235. Liu, J., et al., Automatic measurement for dimensional changes of woven fabrics based on texture. *Measurement Science and Technology*, 2014. **25**(1): 015602.
236. Candan, C., U. Nergis, and Y. İridağ, Performance of open-end and ring spun yarns in weft knitted fabrics. *Textile Research Journal*, 2000. **70**(2): 177–181.
237. Candan, C. and L. Önal, Dimensional, pilling, and abrasion properties of weft knits made from open-end and ring spun yarns. *Textile Research Journal*, 2002. **72**(2): 164–169.
238. Chen, Q., et al., An analysis of the felting shrinkage of plain knitted wool fabrics. *Textile Research Journal*, 2004. **74**(5): 399–404.
239. Dobilaite, V. and M. Juciene, The influence of mechanical properties of sewing threads on seam pucker. *International Journal of Clothing Science and Technology*, 2006. **18**(5): 335–345.
240. Cardamone, J.M., Activated peroxide for enzymatic control of wool shrinkage part II: Wool and other fiber-type fabrics. *Textile Research Journal*, 2006. **76**(2): 109–115.
241. Gore, S., et al., Standardizing a pre-treatment cleaning procedure and effects of application on apparel fabrics. *Textile Research Journal*, 2006. **76**(6): 455–464.
242. Malčiauskienė, E., Ž. Rukuižienė, and R. Milašius, Investigation of linen honeycomb weave fabric shrinkage after laundering in pure water. *Fibres & Textiles in Eastern Europe*, 2008. **16**(6): 71.
243. Heap, S., et al., Prediction of finished weight and shrinkage of cotton knits: The starfish project, part I: Introduction and general overview. *Textile Research Journal*, 1983. **53**: 109–119.
244. Heap, S., et al., Prediction of finished relaxed dimensions of cotton knits. *Textile Research Journal*, 1985. **55**(4): 211.

245. Fletcher, H.M. and S.H. Roberts, The geometry of plain and rib knit cotton fabrics and its relation to shrinkage in laundering. *Textile Research Journal*, 1952. **22**(2): 84–88.

246. Collins, G., Fundamental principles that govern the shrinkage of cotton goods by washing. *Journal of the Textile Institute Proceedings*, 1939. **30**(3): 46–61.

247. Hearle, J., The nature of setting, in *Setting of Fibres and Fabrics*. 1971, Merrow Publishing Co.: Watford.

248. Pierce, F., The geometry of cloth structure. *Journal of Textile Institute*, 1937. **28**: 45–96.

249. Mehta, P., *An Introduction to Quality Control for the Apparel Industry*. 1992, Quality Press: New York.

250. Özdil, N., Stretch and bagging properties of denim fabrics containing different rates of elastane. *Fibres & Textiles in Eastern Europe*, 2008. **16**(1): 66.

251. Nayak, R., et al., Handle and comfort properties of polyester/viscose suiting fabrics. *Man-Made Textiles in India*, 2007. **50**(8): 288–292.

252. Chatterjee, K., D. Das, and R. Nayak, Study of handle and comfort properties of poly-khadi, handloom and powerloom fabrics. *Man-Made Textiles in India*, 2011. **39**(10): 351.

253. Nayak, R., et al., Comfort properties of suiting fabrics. *Indian Journal of Fibre & Textile Research*, 2009. **34**: 122–128.

254. Uzun, M., Effect of ultrasonic laundering on thermophysiological properties of knitted fabrics. *Fibers and Polymers*, 2013. **14**(10): 1714–1721.

255. Vaeck, S., Chemical and Mechanical Wear of Cotton Fabric in Laundering. *Journal of the Society of Dyers and Colourists*, 1966. **82**(10): 374–379.

256. Parisot, A. and A. Fresco, *Bulletin of Institut Textile de France*, 1954. **48**: 7.

257. Gagliardi, D., A. Wehner, and R. Cicione, Improved durable-press cottons produced by conventional pad-dry-cure procedures using pairs of monofunctional and difunctional swelling reactants 1. *Textile Research Journal*, 1968. **38**(4): 426.

258. Handu, J., K. Sreenivas, and S. Ranganathan, Chemical and mechanical damage in service wear of cotton apparel fabrics. *Textile Research Journal*, 1967. **37**(11): 997.

259. Rollins, M., et al., Abrasion phenomena in durable-press cotton fabrics: A microscopical view. *Textile Research Journal*, 1970. **40**(10): 903–916.

260. Daroux, F., et al., Effect of laundering on blunt force impact damage in fabrics. *Forensic Science International*, 2010. **197**(1): 21–29.

261. Taupin, J.M., F. Adolf, and J. Robertson, *Examination of Damage to Textiles*. Forensic Examination of Fibres, 1999, pp. 65–87, Vol. 2.

262. Klepp, I., Clothes and cleanliness: Why we still spend as much time on laundry. *Ethnologia Scandinavica*, 2003. **33**: 61–73.

263. Fisher, T., et al., *Public Understanding of Sustainable Clothing: A Report to the Department for Environment, Food and Rural Affairs*. 2008, DEFRA: London.

264. Kerr, K.-L., S.J. Rosero, and R.L. Doty, Odors and the perception of hygiene. *Perceptual and Motor Skills*, 2005. **100**(1): 135–141.

265. Ashenburg, K., *The Dirt on Clean: An Unsanitized History*. 2010: Random House LLC. New York City, New York.

266. Uitdenbogerd, D.E., *Energy and Households: The Acceptance of Energy Reduction Options in Relation to the Performance and Organisation of Household Activities*. 2007, Wageningen University: Wageningen.

267. Fletcher, K.T. and P.A. Goggin, The dominant stances on ecodesign: A critique. *Design Issues*, 2001. **17**(3): 15–25.

268. Beal, C.D., E. Bertone, and R.A. Stewart, Evaluating the energy and carbon reductions resulting from resource-efficient household stock. *Energy and Buildings*, 2012. **55**: 422–432.

268a. Witt, C.S. and J. Warden, Can home laundries stop the spread of bacteria in clothing? *Textile Chemist & Colorist*, 1971. **3**(7).

269. Mac Namara, C., et al., Dynamics of textile motion in a front-loading domestic washing machine. *Chemical Engineering Science*, 2012. **75**: 14–27.

270. Murray, A. and E. Ho. New motion control architecture simplifies washing machine motor control system development, in *Industry Applications Conference, 2006. 41st IAS Annual Meeting. Conference Record of the 2006 IEEE*. 2006: IEEE, Florida.

271. Sérgio, A., et al., The design of a washing machine prototype. *Materials & Design*, 2003. **24**(5): 331–338.

272. Blokker, E., J. Vreeburg, and J. Van Dijk, Simulating residential water demand with a stochastic end-use model. *Journal of Water Resources Planning and Management*, 2009. 136(1): 19–26.

273. Laitala, K. and K. Vereide, *Washing Machines' Program Selections and Energy Use*. 2010, National Institute for Consumer Research: Oslo, Norway.

274. Gurudatt, K., V.M. Nadkarni, and K.C. Khilar, A study on drying of textile substrates and a new concept for the enhancement of drying rate. *The Journal of The Textile Institute*, 2010. **101**(7): 635–644.

275. Chatterjee, K., et al., Care labelling of apparels. *Indian Textile Journal*, 2006:116(5): 55–60.

276. Davis, L.L., Consumer use of label information in ratings of clothing quality and clothing fashionability. *Clothing and Textiles Research Journal*, 1987. **6**(1): 8–14.

277. Mehta, P.V. and S.K. Bhardwaj, *Managing Quality in the Apparel Industry*. 1998, New Age International: New Delhi, India.

278. Feltham, T. and L. Martin, Apparel care labels: Understanding consumers' use of information. *Marketing*, 2006. **27**(3): 231–244.

279. Nayak, R. and R. Padhye, *Garment Manufacturing Technology*. 2015, Woodhead Publishing: Cambridge, UK.

280. Kadolph, S., *Quality Assurance for Textiles and Apparel*. 2nd ed. 2007, Fairchild Publications, Inc: New York.

281. Duncan Kariuki, N., et al., Clothing standards compliance assessment: The modeling and application of clothing standards compliance index. *International Journal of Clothing Science and Technology*, 2014. **26**(5): 377–394.

282. Koester, A.W. and J.K. May, Profiles of adolescents' clothing practices: Purchase, daily selection, and care. *Adolescence*, 1985. 20(77): 97.

283. Heisey, F.L., Perceived quality and predicted price: Use of the minimum information environment in evaluating apparel. *Clothing and Textiles Research Journal*, 1990. **8**(4): 22–28.

284. Holmlund, M., A. Hagman, and P. Polsa, An exploration of how mature women buy clothing: Empirical insights and a model. *Journal of Fashion Marketing and Management*, 2011. **15**(1): 108–122.

285. Huddleston, P., N.L. Cassill, and L.K. Hamilton, Apparel selection criteria as predictors of brand orientation. *Clothing and Textiles Research Journal*, 1993. **12**(1): 51–56.

286. Yan, R.-N., J. Yurchisin, and K. Watchravesringkan, Use of care labels: Linking need for cognition with consumer confidence and perceived risk. *Journal of Fashion Marketing and Management*, 2008. **12**(4): 532–544.

287. GINETEX - The international association for textile care labelling. 2014; Available from: http://www.ginetex.net/ginetex.

288. ISO, *ISO 3758: 91 Textiles – Care Labelling Code Using Symbols*. 2012, International Organization for Standardization, Geneva, Switzerland.

289. Nayak, R., et al., RFID: Tagging the new era. *Man-Made Textiles in India*, 2007. **50**(5): 174–177.

290. Nayak, R. and R. Padhye, Introduction: The apparel industry, in *Garment Manufacturing Technology*, R. Nayak and R. Padhye, Editors. 2014, Woodhead Publishing: Cambridge, UK, pp. 1–18.

291. Nayak, R., et al., The role of mass customisation in the apparel industry. *International Journal of Fashion Design, Technology and Education*, 2015. **8**(2): 162–172.

292. Nayak, R., et al., RFID: Tagging the new era. *Man-Made Textiles in India*, 2007. **50**(5): 174–177.

293. Nayak, R., et al., RFID in textile and clothing manufacturing: Technology and challenges. *Fashion and Textiles*, 2015. **2**(1): 1–16.

294. Vezzoli, C., Clothing care in the sustainable household. *Partnership and Leadership*, Seventh International Conference of Greening of Industry Network, Rome, 1998, pp. 1–32.

295. Gururajan, A., E.F. Hequet, and H. Sari-Sarraf, Objective evaluation of soil release in fabrics. *Textile Research Journal*, 2008. **78**(9): 782–795.

296. Petrick, L.M., D.N. Hild, and S.K. Obendorf, Observations of soiling of nylon 66 carpet fibers. *Textile Research Journal*, 2006. **76**(3): 253–260.

297. Holme, I., Innovative technologies for high performance textiles. *Coloration Technology*, 2007. **123**(2): 59–73.

298. Sercombe, J.K., et al., The vertical distribution of house dust mite allergen in carpet and the effect of dry vacuum cleaning. *International Journal of Hygiene and Environmental Health*, 2007. **210**(1): 43–50.

299. Corsi, R.L., J.A. Siegel, and C. Chiang, Particle resuspension during the use of vacuum cleaners on residential carpet. *Journal of Occupational and Environmental Hygiene*, 2008. **5**(4): 232–238.

300. Yiin, L.-M., et al., Cleaning efficacy of high-efficiency particulate air-filtered vacuuming and "dry steam" cleaning on carpet. *Journal of Occupational and Environmental Hygiene*, 2007. **5**(2): 94–99.

301. Ohl, M., et al., Hospital privacy curtains are frequently and rapidly contaminated with potentially pathogenic bacteria. *American Journal of Infection Control*, 2012. **40**(10): 904–906.

302. Vanlerberghe, V., et al., Residual insecticidal activity of long-lasting deltamethrin-treated curtains after 1 year of household use for dengue control. *Tropical Medicine & International Health*, 2010. **15**(9): 1067–1071.

303. Khippal, A. and P. Sharma, Associated problems of consumers after curtains and upholstery purchase. *Asian Journal of Home Science*, 2010. **5**(2): 286–289.

304. Tovey, E.R. and L.G. McDonald, A simple washing procedure with eucalyptus oil for controlling house dust mites and their allergens in clothing and bedding. *Journal of Allergy and Clinical Immunology*, 1997. **100**(4): 464–466.

305. Arlian, L.G., D.L. Vyszenski-Moher, and M.S. Morgan, Mite and mite allergen removal during machine washing of laundry. *Journal of Allergy and Clinical Immunology*, 2003. **111**(6): 1269–1273.

306. Deng, B., et al., Laundering durability of superhydrophobic cotton fabric. *Advanced Materials*, 2010. **22**(48): 5473–5477.

307. Laitala, K., et al., *Consumers' Wool Wash Habits-and Opportunities to Improve Them*. 2011, National Institute for Consumer Research: Oslo, Norway.

308. Madsen, J., et al., Mapping of evidence on sustainable development impacts that occur in life cycles of clothing: A report to the department for environment, food and rural affairs, in *Environmental Resources Management (ERM) Ltd. Defra, London*. 2007, A Report to DEFRA.

309. Saouter, E., et al., The effect of compact formulations on the environmental profile of Northern European granular laundry detergents. Part II: Life Cycle assessment. *International Journal of Life Cycle Assessment*, 2002. **7**(1): 27–38.

310. Terpstra, P.M., Domestic and institutional hygiene in relation to sustainability. Historical, social and environmental implications. *International Biodeterioration & Biodegradation*, 1998. **41**(3): 169–175.

311. Halvorsen-Gunnarsen, J., Laundry detergents for professional use, *Norske vaskeriers kvalitetstilsyn (Norwegian Laundries Quality Control Authority)*. 2011, Oslo, Norway.

312. Simpson, W. and G. Crawshaw, *Wool: Science and Technology*. 2002, CRC Press, Woodhead Publishing: Cambridge, UK.

313. King, G., Some frictional properties of wool and nylon fibres. *Journal of the Textile Institute Transactions*, 1950. **41**(4): T135–T144.

314. Bohm, L., The frictional properties of wool fibres in relation to felting. *Journal of the Society of Dyers and Colourists*, 1945. **61**(11): 278–283.

315. Martin, A., Observations on the theory of felting. *Journal of the Society of Dyers and Colourists*, 1944. **60**(12): 325–328.

316. Lacasse, K. and W. Baumann, *Textile Chemicals: Environmental Data and Facts*. 2004: Springer.

317. Ukponmwan, J., A. Mukhopadhyay, and K. Chatterjee, Pilling, in *Textile Progress*. 1998, Taylor & Francis, London, pp. 1–57.

318. Sharma, I., et al., Critical appraisal of pilling on polyester worsted fabric. *Indian Journal of Fibre & Textile Research*, 1996. **21**(2): 122–126.

319. Hu, C. and Y. Jin, Wash-and-wear finishing of silk fabrics with a water-soluble polyurethane. *Textile Research Journal*, 2002. **72**(11): 1009–1012.

320. Van Amber, R.R., B.E. Niven, and C.A. Wilson, Effects of laundering and water temperature on the properties of silk and silk-blend knitted fabrics. *Textile Research Journal*, 2010. 80(15): 1557–1568.

321. Ma, M., et al., Effects of ultrasonic laundering on the properties of silk fabrics. *Textile Research Journal*, 2014. **84**(20): 2166–2174.

322. Yang, H. and C.Q. Yang, Durable flame retardant finishing of the nylon/cotton blend fabric using a hydroxyl-functional organophosphorus oligomer. *Polymer Degradation and Stability*, 2005. **88**(3): 363–370.

323. Gericke, A., L. Viljoen, and R. de Bruin, Compatibility of cotton/nylon and cotton/polyester warp-knit terry towelling with industrial laundering procedures. 2005.

324. Venkatesh, G., et al., A study of the soiling of textiles and development of anti-soiling and soil release finishes: A review. *Textile Research Journal*, 1974. **44**(5): 352–362.

325. Nayak, R., et al., Care and maintenance of textile products: wWith special emphasis on protective textiles. *Textile Progress*, 2015. **47**.

326. Houshyar, S., et al., Deterioration of polyaramid and polybenzimidazole woven fabrics after ultraviolet irradiation. *Journal of Applied Polymer Science*, 2016. **133**(9): 1–7, 10.1002/app.43073.

327. Rushton, C., et al., The distribution and significance of amylase-containing stains on clothing. *Journal of the Forensic Science Society*, 1979. **19**(1): 53–58.

328. Hunt, D. and O. Makinson, Removal of plaque disclosing stains from clothing. *Australian Dental Journal*, 1984. **29**(1): 5–9.

329. Rowe, H.D., Biotechnology in the textile/clothing industry—A review. *Journal of Consumer Studies & Home Economics*, 1999. **23**(1): 53–61.

330. Bevan, G., Fabric washing in Western Europe. *Review of Progress in Coloration and Related Topics*, 1997. **27**(1): 1–4.

331. Flores, A.C., et al., Clinical efficacy of 5% sodium hypochlorite for removal of stains caused by dental fluorosis. *Journal of Clinical Pediatric Dentistry*, 2009. **33**(3): 187–192.

332. Sansone, E.B. and L.A. Jonas, The effect of exposure to daylight and dark storage on protective clothing material permeability. *The American Industrial Hygiene Association Journal*, 1981. **42**(11): 841–843.

333. Coombs, R. and B. Dodd, Possible application of the principle of mixed agglutination in the identification of blood stains. *Medicine, Science and the Law*, 1961. **1**(4): 359–377.

334. Reeves, W.A., T.A. Summers, and R.M. Reinhardt, Soiling, Staining, and Yellowing Characteristics of Fabrics Treated With Resin or Formaldehyde. *Textile Research Journal*, 1980. **50**(12): 711–717.

335. Fleming, A., *Blood Stains in Criminal Trials*. 1861: WS Editions, New Haven, CT.

336. Harwood, F.C., Modern detergency. *Journal of the Textile Institute Proceedings*, 1953. **44**(3): P105–P113.

337. Neiditch, O., K. Mills, and G. Gladstone, The stain removal index (SRI): A new reflectometer method for measuring and reporting stain removal effectiveness. *Journal of the American Oil Chemists' Society*, 1980. **57**(12): 426–429.

338. Nayak, R., et al., Airbags, in *Textile Progress*. 2013, Taylor & Francsis: UK, pp. 209–301.

339. Laitala, K., C. Boks, and I.G. Klepp, Potential for environmental improvements in laundering. *International Journal of Consumer Studies*, 2011. **35**: 254–264.

340. Zoller, U., *Handbook of Detergents: Environmental Impact*. 2004, CRC Press: Boca Raton, FL.

341. Liwarska-Bizukojc, E., et al., Acute toxicity and genotoxicity of five selected anionic and nonionic surfactants. *Chemosphere*, 2005. **58**(9): 1249–1253.

342. McCoy, M., Greener cleaners. *Chemical & Engineering News*, 2008. **86**(3): 15–23.

343. Harris, R., S.H. Jureller, and J.L. Kerschner, Method of dry cleaning fabrics using densified liquid carbon dioxide. US Patent 5,683,473. 1997, US Patents.

344. Preston, A.D. and J.R. Turner, Carbon dioxide dry cleaning system. US Patent 5,904,737. 1999, US Patents.

345. Bae-Lee, M., et al., Dry cleaning system using densified carbon dioxide and a surfactant adjunct. US 5683977 A. 1997, US Patents.

346. Rosio, L.R. and S.H. Shore, Apparatus and method for controlling the use of carbon dioxide in dry cleaning clothes. US Patent 5,970,554. 1999, US Patents.

347. Murphy, D.S., Dry cleaning system using densified carbon dioxide and a surfactant adjunct. US Patent 5,977,045. 1999, Google Patents.

348. Van Roosmalen, M., et al., Dry-cleaning with high-pressure carbon dioxide-the influence of mechanical action on washing results. *The Journal of Supercritical Fluids*, 2003. **27**(1): 97–108.

349. Fijan, S., R. Fijan, and S. Šostar-Turk, Implementing sustainable laundering procedures for textiles in a commercial laundry and thus decreasing waste-water burden. *Journal of Cleaner Production*, 2008. **16**(12): 1258–1263.

350. Berth, P., Recent developments in the field of inorganic builders. *Journal of the American Oil Chemists' Society*, 1978. **55**(1): 52–57.

351. Rice, R.G., et al., Microbiological benefits of ozone in laundering systems. *Ozone: Science & Engineering*, 2009. **31**(5): 357–368.

352. Gallego-Juarez, J.A., High-power ultrasonic processing: Recent developments and prospective advances. *Physics Procedia*, 2010. **3**(1): 35–47.

353. Gallego-Juarez, J.A., et al., Ultrasonic system for continuous washing of textiles in liquid layers. *Ultrasonics Sonochemistry*, 2010. **17**(1): 234–238.

354. Canoğlu, S., B. Gültekin, and S. Yükseloğlu, Effect of ultrasonic energy in washing of medical surgery gowns. *Ultrasonics*, 2004. **42**(1): 113–119.

355. Warmoeskerken, M., et al., Laundry process intensification by ultrasound. *Colloids and Surfaces A: Physicochemical and Engineering Aspects*, 2002. **210**(2): 277–285.

356. Moholkar, V.S., et al., Mechanism of mass-transfer enhancement in textiles by ultrasound. *AIChE Journal*, 2004. **50**(1): 58–64.

357. Gotoh, K. and K. Harayama, Application of ultrasound to textiles washing in aqueous solutions. *Ultrasonics Sonochemistry*, 2013. **20**(2): 747–753.

358. Hurren, C.J., *A Study into the Ultrasonic Cleaning of Wool*. 2010, Deakin University: Geelong, Australia.

359. Gaete-Garretón, L., et al., On the onset of transient cavitation in gassy liquids. *The Journal of the Acoustical Society of America*, 1997. **101**(5): 2536–2540.

360. Smulders, E., et al., *Laundry Detergents*. 2007, Wiley-VCH Verlag GmbH & Co. KGaA: Germany.

361. Flick, E.W., *Advanced Cleaning Product Formulations*. Vol. 2. 2013, Elsevier: Amsterdam, Netherlands.

362. Case, F., Silicones in fabric care. *Journal of Surfactants and Detergents*, 2006. **9**(4): 303.

363. Zhang, X. and B. Han, Cleaning using CO_2-based solvents. *Clean: Soil, Air, Water*, 2007. **35**(3): 223–229.

364. Materna, B.L., Occupational exposure to perchloroethylene in the dry cleaning industry. *The American Industrial Hygiene Association Journal*, 1985. **46**(5): 268–273.

365. Kyyrönen, P., et al., Spontaneous abortions and congenital malformations among women exposed to tetrachloroethylene in dry cleaning. *Journal of Epidemiology and Community Health*, 1989. **43**(4): 346–351.

366. Eskenazi, B., et al., A study of the effect of perchloroethylene exposure on the reproductive outcomes of wives of dry-cleaning workers. *American Journal of Industrial Medicine*, 1991. **20**(5): 593–600.

367. Hellweg, S., et al., Confronting workplace exposure to chemicals with LCA: Examples of trichloroethylene and perchloroethylene in metal degreasing and dry cleaning. *Environmental Science & Technology*, 2005. **39**(19): 7741–7748.

368. Duh, R.-W. and N.R. Asal, Mortality among laundry and dry cleaning workers in Oklahoma. *American Journal of Public Health*, 1984. **74**(11): 1278–1280.

369. Lauwerys, R., et al., Health surveillance of workers exposed to tetrachloroethylene in dry-cleaning shops. *International Archives of Occupational and Environmental Health*, 1983. **52**(1): 69–77.

370. Brown, D.P. and S.D. Kaplan, Retrospective cohort mortality study of dry cleaner workers using perchloroethylene. *Journal of Occupational and Environmental Medicine*, 1987. **29**(6): 535–541.

371. Ruder, A.M., E.M. Ward, and D.P. Brown, Mortality in dry-cleaning workers: An update. *American Journal of Industrial Medicine*, 2001. **39**(2): 121–132.

372. Kawauchi, T. and K. Nishiyama, Residual tetrachloroethylene in dry-cleaned clothes. *Environmental Research*, 1989. **48**(2): 296–301.

373. Sherlach, K.S., et al., Quantification of perchloroethylene residues in dry-cleaned fabrics. *Environmental Toxicology and Chemistry*, 2011. **30**(11): 2481–2487.

374. Nayak, R., R. Padhye, and S. Fergusson, Identification of natural textile fibres, in *Handbook of Natural Fibres, Volume 1—Types, Properties and Factors Affecting Breeding and Cultivation*, R. Kozlowski, Editor. 2012, Woodhead Publishing: Cambridge, UK, p. 314.

375. Gloxhuber, C. and K. Klunstler, *Anionic Surfactants: Biochemistry, Toxicology, Dermatology*. Vol. 43. 1992, Marcel Dekker, Inc.: New York.

376. Bircher, A., Cutaneous immediate-type reactions to textiles, in *Textiles and the Skin*, P. Elsner, K. Hatch, and W. Wigger-Alberti, Editors. 2004, Karger Publishers: Berlin, Germany, pp. 166–170.

377. Gerba, C.P. and D. Kennedy, Enteric virus survival during household laundering and impact of disinfection with sodium hypochlorite. *Applied and Environmental Microbiology*, 2007. **73**(14): 4425–4428.

378. Callaghan, I., Bacterial contamination of nurses' uniforms: A study. *Nursing Standard*, 1998. **13**(1): 37–42.

379. Treakle, A.M., et al., Bacterial contamination of health care workers' white coats. *American Journal of Infection Control*, 2009. **37**(2): 101–105.

380. Barrett, T.W. and G.J. Moran, Update on emerging infections: News from the Centers for Disease Control and Prevention. *Annals of Emergency Medicine*, 2004. **43**(1): 45–47.

381. McCoy, M., Going green. *Chemical and Engineering News*, 2007. **85**(13): 9.

382. Park, J.H. and L. Stoel, Apparel shopping on the internet: Information availability on US apparel merchant web sites. *Journal of Fashion Marketing and Management*, 2002. **6**(2): 158–176.

383. Kim, J.-H. and S.J. Lennon, Information available on a web site: Effects on consumers' shopping outcomes. *Journal of Fashion Marketing and Management*, 2010. **14**(2): 247–262.

384. Xu, Y. and V.A. Paulins, College students' attitudes toward shopping online for apparel products: Exploring a rural versus urban campus. *Journal of Fashion Marketing and Management*, 2005. **9**(4): 420–433.

385. Nayak, R., et al., Evaluation of functional finishes–An overview. *Man-Made Textiles in India*, 2008. **51**(4): 131–135.

386. Ryom, N., Adjusting laundry detergents to modern needs with enzymes. *Rivista Italiana delle Sostanze Grasse*, 2003. **80**(5): 313–316.

387. Van Hoof, G., D. Schowanek, and T.C. Feijtel, Comparative life-cycle assessment of laundry detergent formulations in the UK. Part I: Environmental fingerprint of five detergent formulations in 2001. *Tenside Surfactants Detergents,* 2003. **40**(5): 266–275.

388. Kefgen, M. and Touchie-Specht, P., *Individuality in Clothing Selection and Personal Appearance, A Guide for the Consumer,* 3rd ed. 1981, Macmillan Publishing Co., Inc.: New York.

389. NFPA 2112-2012. Standard on flame-resistant clothing for protection of industrial personnel against short-duration thermal exposures from fire.

Index

Page numbers followed by *f* indicate figures; those followed by *t* indicate tables.

Proper care and maintenance of textile materials is essential in prolonging their durability and appearance. This book describes methods of care and maintenance for textile products, focusing on types of laundering and dry-cleaning processes, chemicals, and equipment, while considering the environmental impacts of these procedures and green cleaning approaches. It details care labelling of garments, including electronic care labelling and instructions for different specialty textiles. Factors such as pilling, abrasion, snagging, color fading, and dimensional change are discussed. This book also emphasizes care and maintenance of textiles used for protection from fire, bullets, cold weather, and chemicals.

- Details garment care and maintenance

- Describes care and maintenance of various types of protective clothing

- Illustrates environmental impacts of care and maintenance and greener approaches

- Covers care and maintenance of specialty textiles

- Discusses care labelling and care labelling systems

A Textile Institute Professional Publication

The Textile Institute

K43574

CRC Press
Taylor & Francis Group
an **informa** business

6000 Broken Sound Parkway, NW
Suite 300, Boca Raton, FL 33487
711 Third Avenue
New York, NY 10017
2 Park Square, Milton Park
Abingdon, Oxon OX14 4RN, UK

ISBN: 978-1-138-56581-4

9 781138 565814

90000

www.crcpress.com